Eastern a[...]

by Stephen K. Donovan, Trevor A. Jackson,
Harold L. Dixon and Eamon N. Doyle

Edited by
J. T. Greensmith

© THE GEOLOGISTS'
ASSOCIATION
1995

ISBN 0-900717-77-7

Eastern and Central Jamaica

CONTENTS

PREFACE	1
GEOLOGICAL HISTORY	1
ITINERARIES	
Round Hill, Clarendon	7
Discovery Bay and Duncans	12
Southern Wagwater Belt	17
Northeast Coast	23
Southeast Coast	27
Bauxite Mines	32
St. Mary Coast	33
Central Inlier	36
Brown's Town, St. Ann	40
Above Rocks and Benbow Inliers	43
ACKNOWLEDGEMENTS	47
REFERENCES	48
GLOSSARY	60

Eastern and Central Jamaica

PREFACE

This field guide offers ten excursions which introduce the stratigraphy and structure of Jamaica to all people interested in geology. They are a consequence of a Field Meeting of the Geologists' Association held on the island in January 1993 and led by members of the staff of the Department of Geology, University of the West Indies, Mona, Kingston 7, Jamaica. Although the excursions start from Kingston, many can be undertaken from other centres, such as Port Antonio, Ocho Rios and Discovery Bay.

Some comments on fieldwork in Jamaica may be useful. Grid references used herein refer to the 1:50,000 topographic (metric edition) maps, which can be obtained from the Survey Department, $23^{1}/_{2}$ Charles Street, P.O. Box 493, Kingston, Jamaica or Edward Stanford Ltd., 12/14 Long Acre, London, WC2E 9LP, England. Large scale road maps are available from many service stations and bookshops on the island. 1:50,000 geological maps (provisional series; coverage of eastern Jamaica is incomplete) may be obtained from the Geological Survey Department, Hope Gardens, Kingston 6, Jamaica. If you require further information on specific aspects of Jamaican geology or fieldwork on the island, then do not hesitate to contact the Department of Geology at UWI.

Local transport in Jamaica is often difficult. The visitor to the island who is intending to undertake fieldwork is recommended to hire a car; however, costs of hire cars in Jamaica are often high. The only other reliable form of transport is by bus. While cheap, buses are grossly overcrowded and tend to stick to the main routes, particularly in the country.

Jamaicans are generally friendly people. The field geologist will find that in the country local people will be courteous, interested in what you are doing, often keen to help and may even be able to direct you to other nearby rock exposures. However, in common with cities elsewhere, the main centres of population (Kingston, Ocho Rios and, in particular, Montego Bay) attract a certain number of ne'er-do-wells, so it is sensible to exert caution in these areas.

GEOLOGICAL HISTORY

This account of the geological history of Jamaica is slightly expanded after Donovan (1993a). The principal geological features of the island are illustrated in Figures 1 to 3.

The oldest rocks exposed in Jamaica are of Cretaceous age. The succession is dominated by andesitic volcanic rocks with associated limestones and represents an island arc system (Draper, 1987). In the early

Eastern and Central Jamaica

Cretaceous the western end was the site of a back-arc basin, the centre of the island was located on a volcanic arc, and the eastern end was a region of subduction-related sedimentation and metamorphism. In the late Cretaceous the island arc moved further east.

The oldest Cretaceous rocks are Barremian (early Cretaceous) in age and are exposed in the Benbow Inlier (Figure 2) in the eastern Clarendon Block. The sequence consists of volcanic lavas and volcaniclastics that were erupted in a subaerial to shallow nearshore environment. Associated limestones were deposited further offshore (Burke et al., 1968). Volcanic centres were concentrated in the southern and eastern parts of the Clarendon Block (Figure 3), where volcanicity continued until the Campanian (late Cretaceous). These volcanic deposits are interbedded with shales and limestones. In western Jamaica, volcanic rocks and shales were accumulating in a basinal environment. Pre-Campanian rocks in eastern Jamaica are now metamorphosed into blueschists, greenschists and amphibolites (Draper et al., 1976; Draper, 1978) and were originally deposited in a geosynclinal (fore-arc) setting, with a subduction zone to the southeast sloping beneath Jamaica (Draper, 1987).

Figure 1: Geological map of Jamaica, showing the principal stratigraphical units (after Donovan, 1993a, fig. 1; based on Geological Survey of Jamaica, 1958; see also McFarlane, 1977a).
Key: B=Blue Mountain Inlier; C=Central Inlier. Ages of principal Cenozoic units: granodiorite=Upper Cretaceous to Palaeocene; Wagwater Formation, Newcastle Volcanics=Palaeocene; Richmond Formation=Palaeocene to Lower Eocene; Yellow Limestone Group=Lower to Middle Eocene; White Limestone Supergroup=Middle Eocene to Upper Miocene; Coastal Group=Upper Miocene to Quaternary; alluvium=Quaternary.
The inset map shows the position of Jamaica in the Caribbean; J=Jamaica; C=Cuba; H=Hispaniola (Haiti+Dominican Republic); PR=Puerto Rico; LA=Lesser Antilles; T=Trinidad; V=Venezuela; Co=Colombia.

Eastern and Central Jamaica

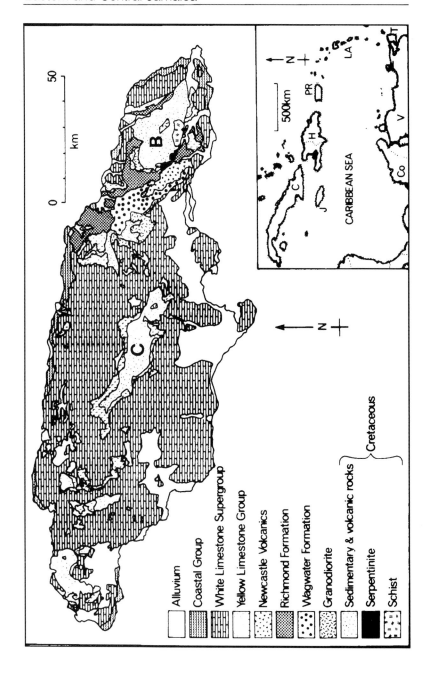

Eastern and Central Jamaica

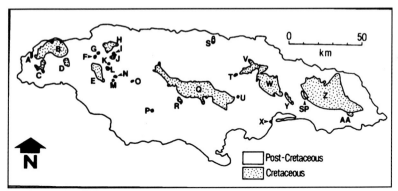

Figure 2: The Cretaceous inliers of Jamaica (after Donovan & Bowen, 1989) Key: A=Green Island; B=Lucea; C=Jerusalem Mountain; D=Grange; E=Marchmont; F=Seven Rivers; G=Mafoota; H=Sunderland; I=Calton Hill; J=Maldon; K=Garlands; L=Mocho-Sweetwater; M=Barracks River; N=Elderslie; O=Aberdeen; P=Nottingham; Q=Central; R=Banana Ground; S=St. Ann; T=Mount Diablo; U=Giblatore; V=Benbow-Guys Hill; W=Above Rocks; X=Lazaretto; Y=Jacks Hill; Z=Blue Mountain; AA=Sunning Hill. SP is the so-called St Peter's Inlier, which is probably a suite of deformed Tertiary rocks (Jackson, 1986).

Volcanism in Jamaica had largely ceased by the early Maastrichtian (late Cretaceous). Volcanic rocks in central Jamaica were eroded to form red beds, which are overlain by limestones interbedded with clastic sedimentary rocks following marine transgression. In southeast Jamaica, at Bath, submarine lavas of equivalent age occur in a sequence of mainly shales and volcaniclastic sedimentary rocks (Wadge *et al.*, 1982).

Renewed andesitic volcanism in the latest Cretaceous led to the cessation of limestone deposition in central and western Jamaica. The western Jamaican succession is a back-arc basin sequence (Grippi, 1980; Schmidt, 1988). The end of the Cretaceous was a time of uplift, with folding and intrusion of granodiorites. These granodiorites intruded Maastrichtian rocks, yet were unroofed by the early Eocene.

Eastern and Central Jamaica

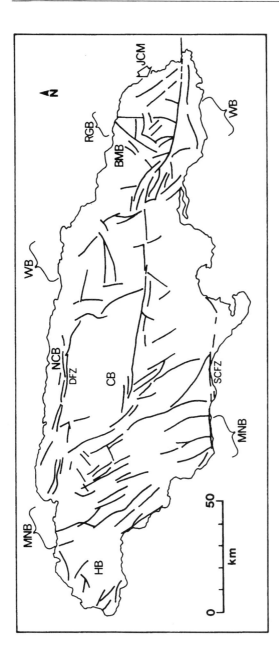

Figure 3: Blocks, belts and major faults of Jamaica (after Donovan, 1993a, fig. 2; modified after Draper, 1987, fig. 1). Key: BMB=Blue Mountain Block; CB=Clarendon Block; DFZ=Duanvale Fault Zone; HB=Hanover Block; JCM=John Crow Mountains Block; MNB=Montpelier-Newmarket Belt; NCB=North Coast Belt; RGB=Rio Grande Belt; SCFZ=South Coast Fault Zone; WB=Wagwater Belt.

Eastern and Central Jamaica

During the earliest Cenozoic the island was uplifted, while the graben (rift valley) of the Wagwater Trough (Figure 3) had opened and was accumulating a thick sequence of clastic sedimentary and volcanic rocks. Much of the island formed an exposed land area, with marine sedimentary rocks only being deposited in the northeast during the Palaeocene (Robinson & Jiang, 1990). At this time a 'block and belt' structure was superimposed by extensional tectonics on the Cretaceous island-arc substratum of Jamaica (Robinson, 1994). A major set of east-west faults truncates a northwest-southeast fault set. The latter set defines three positive blocks separated by graben structures, or belts (Figure 3), which were formed successively in an east-to-west direction from the Palaeocene to middle Eocene (Robinson, 1994).

The early Eocene was a time of marine transgression which flooded western Jamaica, although emergent land areas in the east provided a clastic input to the surrounding basins. Volcanism was waning. Deposition of the Yellow Limestone Group (Figure 1) commenced in the west and spread east. The Yellow Limestone Group comprises a mixture of limestone and clastic sedimentary rock units deposited in fluvial to offshore marine environments (Robinson, 1988). The progressive submergence of the land area (Eva & McFarlane, 1985) is reflected by a reduction of clastic detritus in younger horizons.

Exposed land areas completely disappeared in the late middle Eocene (Eva & McFarlane, 1985). This led to sedimentation dominated by pure limestones of the White Limestone Supergroup (Figure 1) that persisted until the early late Miocene. A range of carbonate environments existed, with limestones being deposited in deep-water, high-energy open-shelf and low-energy lagoonal settings. The White Limestone Supergroup is divided into eleven formations.

Since the end of White Limestone deposition, Jamaica has been undergoing a period of transcurrent tectonics related to left lateral movement on the northern boundary of the Caribbean Plate (Draper, 1987). This active tectonic phase has led to uplift and emergence. Denudation and fluvial transport of materials derived from mid-Tertiary limestones and older rocks led to their re-sedimentation around the periphery of the island as the Coastal Group (Figure 1). Submarine volcanism at Low Layton in northeast Jamaica occurred in the late Miocene (Wadge, 1982). Reef growth was re-established in the late Pliocene to early Pleistocene. Continued tectonism has led to the uplift of these Pleistocene reefs and deeper-water limestones as a series of terraces (Cant, 1972).

Eastern and Central Jamaica

ITINERARIES

The areas visited by the excursions outlined below are shown in Figure 4.

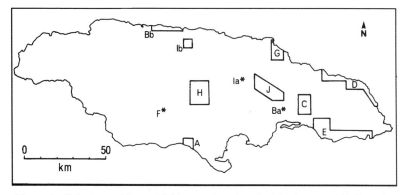

Figure 4: Relative positions of field excursions described in the text. *(A)* Round Hill, parish of Clarendon. *(Ba)* Red Hills viewpoint and cave, St. Andrew. *(Bb)* Discovery Bay, St. Ann to Duncans chalk quarry, Trelawny. *(C)* Southern Wagwater Belt, St. Andrew. *(D)* Northeast coast, Portland. *(E)* Southeast coast, St. Thomas. *(F)* Alcan bauxite works at Kirkvine, Manchester. *(G)* Northern Wagwater Belt, St. Mary. *(H)* Central Inlier, Clarendon. *(Ia)* Mount Diablo red mud lake, St. Catherine. *(Ib)* Brown's Town area, St. Ann. *(J)* Above Rocks and Benbow Inliers, St. Catherine.

ROUND HILL, CLARENDON

Take the main A2 road west from Kingston. Milk River Bath is signposted after the last turnoff for May Pen, about 1½ km before the turning for Four Paths. Alternately, carry on to Toll Gate and take the B12 south (this is the better road). Drive past Milk River Bath and on towards the coast. This stretch of road is poor. Park by the fishing boats on the left, shortly before the beach. Do not drive onto the beach.

Round Hill is a prominent topographic feature on the central south coast of Jamaica in the parish of Clarendon (Donovan *et al.*, 1989). The purpose of this excursion is to examine the Plio-Pleistocene Round Hill Beds, a sequence of marine sedimentary rocks unconformably overlain by a late Quaternary conglomeratic sequence capped by a marine shell bed; the black sand deposit of Farquhar's Beach; the radioactive mineral springs of Milk River Bath; and the spectacular God's Well sinkhole (Figure 5).

Eastern and Central Jamaica

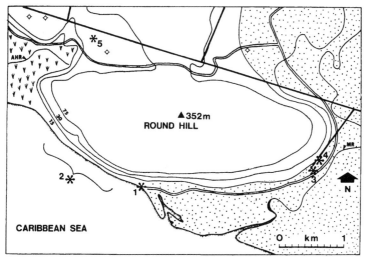

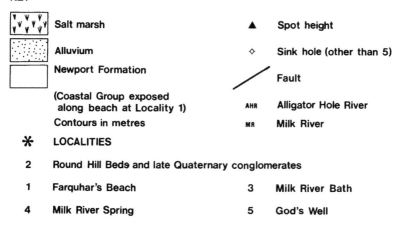

Figure 5: Simplified geological map of the Round Hill area, parish of Clarendon (after Donovan et al., 1989, fig. 1), loosely based on Bateson (1974). The Round Hill Beds and overlying late Quaternary deposits outcrop as a narrow strip adjacent to the coast at Farquhar's Beach, from locality 1 and to the northwest. The Miocene Newport Formation is part of the White Limestone Supergroup.

Eastern and Central Jamaica

Stop 1. Farquhar's Beach

Farquhar's Beach (NGR 094330, continuing to the northwest) is one of several 'black sand' beaches that occur along the south coast of Jamaica. The sand is composed primarily of titano-magnetite and titano-haematite crystals with lesser amounts of feldspar, quartz and calcite (McFarlane, 1977b). Major element contents in the magnetic portion of the black sands include Fe_2O_3 (which ranges from 85.7% to 71.9%), FeO (14.2% to 2.2%) and TiO_2 (16.3% to 8.9%) (Chubb, 1960).

Because the black sands are concentrated at the mouth of the Rio Minho and westwards along the south coast from Farquhar's Beach in Clarendon to Long Acre Point in St. Elizabeth, Chubb (1960) surmised that the sources of the sands were the Cretaceous inliers in the interior of central Jamaica (Figure 2). Erosion and transportation by rivers flowing southwards, accompanied by westerly longshore currents, account for the dispersion of sand along the coast. Wood (1976) noted that sediment samples from Milk River Bay are almost totally composed of clastic material.

Alternatively, McFarlane (1977b) has proposed that the source of the black sands was from an igneous and/or metamorphic outcrop located to the south of the present coastline, but not far from the present location of sand deposits. McFarlane suggested that these rocks were exposed during the low sea stands of the Pleistocene and erosion proceeded mainly by physical weathering. However, this theory is speculative and the available evidence supports the views of Chubb.

Stop 2. Round Hill Beds

The highly fossiliferous succession in the Round Hill Beds (August Town Formation, Coastal Group) at Farquhar's Beach (Figure 5) is sandy limestones and sandstones, with a fossil fauna dominated by benthic molluscs and benthic foraminifers, with less common corals, echinoids, barnacles and trace fossils. Beds dip steeply towards the sea (the shape of the coastline often being controlled by the strike of particularly resistant beds), or are vertical, and the outcrop is incised by a few faults. This coastal exposure was first documented by Duncan and Wall (1865, p. 6, fig. 4), who considered the succession to consist of Miocene sedimentary rocks overlain by a white limestone. Robinson (1968) correctly reinterpreted the structure as a possibly conformable contact between the underlying Miocene Newport Formation (Moneague Group, White Limestone Supergroup) and the younger Round Hill Beds. In turn, the Round Hill Beds are unconformably overlain by a distinctive late Quaternary conglomerate which thickens towards the southeast. Faults in the Round

Eastern and Central Jamaica

Hill Beds are truncated by the angular unconformity. The conglomerate comprises rounded, white limestone clasts (derived from the Newport Formation and including fossils of marine origin) in a red, calcareous matrix which includes calcareous crusts, root casts and terrestrial gastropods that are assignable to taxa still extant on the island. In turn, a bed of marine shells rests unconformably on the terrestrial conglomerates.

The age of the Round Hill Beds is uncertain, but they are at the oldest late Miocene. Robinson (written contribution) considers the sequence to be Pliocene, possibly extending into the Pleistocene. The sedimentary sequence of the Round Hill Beds has yet to be described in detail but it consists of four lithofacies (units A to D; A lies at the base of the sequence) which represent a prograding ('regressive') succession.

Unit A is mainly bedded limestones, which are often nodular. The nodules carry an iron-rich carbonate cement and in at least some examples follow the shape of burrow systems of crustaceans(?) (the trace fossil *Thalassinoides*). Barnacles and oysters are sometimes found on bedding planes, suggesting that the nodular carbonates formed hardground surfaces. The palaeoenvironment is interpreted as open marine and was possibly analogous to that of the carbonate slopes in the northern Bahamas (Mullins *et al.*, 1980). However, this preliminary interpretation must be extended to include a fine grained micritic limestone containing freshwater(?) gastropods underlain by a lignite-rich bed, both of which were discovered by the Geologists' Association party in 1993.

Unit B is a sequence of siliciclastics, particularly planar to cross-bedded sandstones and polymict (mixed origin) pebble conglomerates, with soft sediment slumps(?), minor limestones and occasional oyster beds. Tepee structures, anticline-like in form and indicative of early cementation, expansion and dewatering (Scoffin, 1987, p. 98), are developed near the top. Fossils include various benthic molluscs, the pea-like benthic foraminifer *Sphaerogypsina globulus* (Reuss) (Robinson, 1968, p. 46), bryozoans, rare echinoids (*Clypeaster*) and trace fossils, notably *Dactyloidites ottoi* (Geinitz) (Pickerill *et al.*, 1993a). This unit has an undoubted fluvial aspect, though including fossils of both marine and brackish-water aspect, and may have acted as some sort of barrier between the palaeoenvironments of units A and C.

Robinson (1968, p. 46) noted ". . . Several remarkable beds of oysters occur near the base of the sequence [that is, within unit B], with the oysters in an original position of growth, and with many individual shells reaching 15 inches or more in length." Prescott and Versey (1958, p. 39) considered that these oysters resembled *Ostrea haitiensis* Sowerby, but they are, in fact, *Crassostrea virginica* (Gmelin). One bed of oysters is 3.3 m thick, including

Eastern and Central Jamaica

many individuals in life position, and appears to have been a reef-like shell bank similar to those being formed by the same species in the American Gulf Coast region at the present day (Littlewood & Donovan, 1988; Donovan & Littlewood, 1993b).

Unit C is a sequence of poorly bedded, monotonous, white limestones which abruptly overlie the well-bedded succession of unit B. Fossils in unit C are mainly small, disarticulated oysters and pectinids, with *Spondylus* and *C. virginica* valves near the base. Rare spatangoid fragments have been found higher in the section. While marine in origin, unit C is tentatively interpreted as a possible lagoonal environment, with a restricted shelly benthos, but with the sedimentary sequence highly bioturbated, presumably by a mainly soft-bodied infauna.

Unit D comprises a series of brown-coloured, siliciclastic sedimentary rocks which lack a shelly benthos. This top unit of the sequence is apparently terrestrial (=fluvial?) in origin.

Stop 3. Milk River Bath and Spring

The Milk River Spring discharges through an east-west trending fault at the foot of Round Hill. The main spring provides a supply of water to bathhouses at the Milk River Hotel (3 in Figure 5; NGR 125335). North of the hotel another spring (4 in Figure 5) flows into a shallow well (Royall & Banham, 1981). The springwater at Milk River issues from the Miocene limestone of the Newport Formation, White Limestone Supergroup, close to the contact between the limestone and the alluvium of the Milk River (Fenton, 1981).

The temperature of the water in the bathhouse of the hotel is 33°C, whereas at the well site it is 37°C. The spring waters are known for their high level of radioactivity caused by the presence of radon^{222} (Vincenz, 1959) and are regarded as one of the most radioactive in the world. The springs at Milk River contain up to 16% more chloride than sea water (Royall & Banham, 1981). The Ca/Na ratios of the springs are also higher, which may be due to the increased solubility of calcium carbonate in the spring water.

It is considered that much of the spring water may be derived from sea water entering the South Coast Fault at depth. This water becomes heated and modified in rising along the fault, mixing with groundwater from limestone near the surface. The source of the radon is thought to be at appreciable depth. This was deduced by Versey (1959), who observed that the rate of discharge had a direct relationship to precipitation, but an inverse relationship to the total dissolved solids content of the water and the intensity of radiation.

Eastern and Central Jamaica

Today the Milk River spring is used solely for its therapeutic purposes. Readings of radioactivity of the water in bathhouse number 2 have indicated an average of 7599 counts min^{-1}, with a background of 902 counts min^{-1}; at a small spring at the back of the hotel, the average reading is 1590 counts min^{-1}, with a background of 1407 counts min^{-1} (Donovan et al., 1989, p. 47).

Stop 4. God's Well

God's Well is a spectacular sink hole that occurs in the Newport Formation north of Round Hill. It has been described by Sawkins (1869) and Zans (1960). The major east-west fault which extends along the northern side of Round Hill, plus the northwest-southeast joint systems, have greatly facilitated the karst development of the area (Zans, 1960). God's Well itself was most probably produced by the collapse of a limestone roof over a major underground cavern.

The orifice of the sink hole is oval and measures between 25 and 40 m in diameter. The sink hole has near vertical walls and the distance from the top of the depression to the water level below is approximately 25 m. The water level fluctuates about 0.7 m between wet and dry seasons. God's Well is drained underground to the west, where it issues as a spring at the head of the Alligator Hole River (Zans, 1960).

DISCOVERY BAY AND DUNCANS

Stop 1. Red Hills

Take the Red Hills Road northwest out of Kingston, through Forest Hills and towards Red Hills village. Note that as the road climbs to Red Hills, bauxitic terra rossa soils are seen to overlie the White Limestone Supergroup. Approximately midway between mileposts 9 and 10, a panoramic view of Kingston opens to the south. Points to note include the Liguanea Plain, which is a dormant fan-delta; the Palisadoes, a tombolo, with the town of Port Royal at the end; and the Port Royal Keys, a series of limestone islands seaward of the Palisadoes.

Stop 2. Red Hills bone cave

Continue on the Red Hills Road towards Rock Hall. Stop 2 (Figure 6A; NGR 643573) is about 3.3 km from stop 1. This locality is a cave infill on the south side of the Red Hills Road (Donovan & Gordon, 1989). The cave was presumably exposed during road construction some years ago. Dripstones, fallen boulders and clastic sediment infill are present. The cave is flask-shaped and formed by the solution of limestones of the White

Eastern and Central Jamaica

Limestone Supergroup, with a narrow opening at the apex of the chamber, which is about 10 m high (Figure 7).

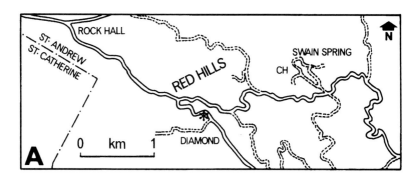

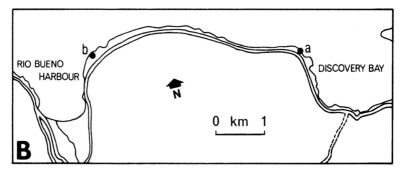

Figure 6: After Donovan & Gordon (1989, fig. 1). (A) Position of the late Quaternary Red Hills bone cave (*), on the south side of the Red Hills Road near Diamond. (B) Localities (see text) exposing the last interglacial Falmouth Formation at Discovery Bay Marine Laboratory (a) and east Rio Bueno Harbour (b). These are localities 5 and 7 of Liddell et al. (1984), respectively.

Eastern and Central Jamaica

Figure 7: The late Quaternary Red Hills bone cave *(after Donovan & Gordon, 1989, fig. 2)*. The vertebrate fauna occurs in the unconsolidated deposits which outcrop behind and beneath the small tree growing in the centre of the figure. Terrestrial gastropods, cemented to the cave wall by dripstone, are arrowed.

Eastern and Central Jamaica

The exceptional feature of this locality is the included terrestrial fauna, presumably of late Pleistocene or Holocene age by comparison with similar deposits from Jamaica and elsewhere in the Caribbean (MacPhee et al., 1989). In common with other Jamaican examples, the dominant elements of the fauna are gastropods (see, for example, Goodfriend & Mitterer, 1988) and vertebrate bones. Savage (1990) recognised amphibians (few, very small bones), reptiles (small lizards), birds (several medium- and small-sized species) and mammals (disarticulated bones). The last includes bat jaws and cranial remains of *Geocapromys brownii*, the Jamaican hutia. Rare millipedes, coated in dripstone, have been recovered and uncommon claws of terrestrial crustaceans are also known.

The level of the terrestrial gastropods, which are cemented as a sheet against the side of the cave by dripstone (Figure 7), reaches a height of 6-7 m, indicating a minimum former level of the sediment surface. However, the present level is about 2 m, much of the unconsolidated sediment having been washed away.

Stop 3. Discovery Bay Marine Laboratory

Drive via Sligoville to Bog Walk and then take the main A1 road north through Ewarton and over Mt. Diablo. At Moneague continue on the A1 (left fork) through Claremont and take the B11 via Bamboo to Brown's Town. Drive north to Discovery Bay, turning east on the A1 to the Marine Laboratory (just past Columbus Park; NGR 060016; Figure 6B, locality a).

Emergent marine terraces occur in some areas behind the narrow island shelf and coastal plain of Jamaica's north coast (Woodley & Robinson, 1977). These terraces are thought to be the consequence of slow tectonic southward tilting of the island, accompanied by block faulting, and eustatic sea-level movements which occurred during the Quaternary (Wright, 1974). Four terraces have been identified on the north coast, all formed of coral reef material associated with shallow-water limestones. The youngest of these terraces (reef terrace 1 of Cant, 1972) was formed on the Falmouth Formation and is of last interglacial age (about 105,000 years old; Ipswichian in terms of the British succession).

The Falmouth Formation (Upper Coastal Group) is composed primarily of back reef deposits, with minor reef crest and near-reef crest sedimentary rocks (Land & Epstein, 1970). They rest unconformably on the eroded surface of the Plio-Pleistocene Hopegate Formation (Upper Coastal Group). Seven biolithofacies have been identified in reconstructing the depositional setting of the Falmouth Formation (Larson, 1983). The rocks are relatively unaltered and approximate to the mineralogy of Recent sediments (Land, 1973).

Eastern and Central Jamaica

The sequence exposed at the west of Discovery Bay, adjacent to the Marine Laboratory, form part of Larson's (1983) molluscan biomicritic wackestone facies (Gordon & Donovan, 1992, fig. 1). It forms a low cliff with a highly karstified surface. The limestone is usually well-lithified, but is occasionally chalky. The fauna comprises small colonial corals, numerous benthic molluscs (Donovan & Littlewood, 1993a, identified 79 species of bivalve, gastropod and scaphopod) and at least three species of echinoids, preserved as disarticulated ossicles (Gordon & Donovan, 1992). The lithology and fauna suggest that a good analogue for this facies is the modern lagoon at Discovery Bay (Boss & Liddell, 1987).

Stop 4. East Rio Bueno Harbour

This locality is about 3 km west of stop 3 and is a low cliff hidden from the road (NGR 023023; Figure 6B, locality b). Travel west on the north coast road and, shortly before the headland on the east side of Rio Bueno harbour, turn right and then keep bearing left. Park at the end of the road which fronts a short row of cottages. Walk through the trees along the cliff top and descend by the path to the beach.

The Falmouth Formation at this stop is poorly lithified, contrasting well with the overlying 'caliche' cap and the underlying, dolomitised Hopegate Formation. The poor cementation of the Falmouth Formation permits the easy recovery of a large diversity of fossils, particularly hermatypic corals and benthic molluscs (almost 100 species identified, including bivalves, gastropods, scaphopods and chitons; Donovan & Littlewood, 1993a), but also crabs (Morris, 1993), echinoid plates (Gordon & Donovan, 1992), ophiuroid ossicles (Donovan *et al.*, 1993a), barnacles and foraminifers. The commonest coral species forming the reef framework include *Porites* spp., *Acropora cervicornis* and *A. palmata* (Liddell *et al.*, 1984).

This deposit represents either a patch reef in a back reef lagoon (Larson, 1983) or part of the shallow fore-reef (Robinson, 1958a; Liddell *et al.*, 1984; Boss & Liddell, 1987). The Falmouth Formation unconformably overlies the Hopegate Formation (late Pliocene), which has been bored by bivalves and clionid sponges. *Siderastrea* sp. and other Falmouth Formation corals also encrust the Hopegate Formation (Liddell *et al.*, 1984). The 'caliche' cap overlying the Falmouth Formation is a 10 to 100 cm thick limestone depleted in magnesium with a fine grained spar cement. It was formed by the filtering of meteoric water downwards (Land, 1973). Usually the contact between the 'caliche' cap and the underlying limestones is sharp.

Eastern and Central Jamaica

Stop 5. West of Rio Bueno

Drive west on the A1 north coast road, through Rio Bueno. Stop 5 (site 9a of Liddell *et al.*, 1984, p. 78) is a convenient point to examine the old sea cliff, inland of the road. Park on the disused road to the left about 0.5 km west of Rio Bueno. This cliff is formed by the Hopegate Formation (Land, 1991), a raised reef formed of well-lithified dolomite and calcite (Liddell *et al.*, 1984, p. 73). Reef terrace 2 (Cant, 1972) is developed on top of the Hopegate Formation. That unit formed a sea cliff at the time of planation of terrace 1 is indicated by the presence of a wave-cut notch. Fossil corals and gastropods, which originally had aragonitic skeletons, are preserved as moulds. Prominent amongst these is the framework of a hermatypic coral, apparently *Acropora cervicornis* (Liddell *et al.*, 1984, fig. 45), which is commonest at Discovery Bay at the present day on the shallow fore-reef below about 10 m depth. This locality may therefore represent part of the fore-reef.

Stop 6. West of Duncans

Continue west on the A1 road and turn into a large, disused chalk quarry on the south side of the road about 5 km west of Duncans (approximate NGR 887020). Chalks of Miocene age are widespread (Ager, 1981, p. 2) and are well-developed in the Montpelier Group of the White Limestone Supergroup, such as at this locality. The presence of nodular cherts and bioclastic layers suggests that this forms part of the Sign Formation (Robinson, 1988, fig. 4). Nodular chert horizons are laterally continuous, but are not very common. A few horizons of greenish clay are bentonites, altered sediment of volcanic origin. These may be related to the ash falls that are assumed to have been weathered to form Jamaica's bauxite deposits (see p.33). The chalks themselves are generally poorly fossiliferous, but loose blocks on the floor of the quarry indicate the presence of bioclastic layers within the sequence. These layers probably indicate storm or earthquake action, moving shelly material from shallower waters. The fauna of these bioclastic limestones includes nummulitid foraminifers, bivalves, corals, and fragments of echinoids and starfishes.

SOUTHERN WAGWATER BELT

This excursion involves a traverse across the south-central portion of the fault-bounded Wagwater Belt, from the Wagwater Fault Zone in the west to the Yallahs Fault Zone (and its northern extension, the Silver Hill Fault Zone) in the east (Figure 8). Contained within the Wagwater Belt is the sequence of Palaeogene sedimentary and volcanic rocks known as the Wagwater Group, which attains a thickness of 3000-6000 m (Green, 1977).

Eastern and Central Jamaica

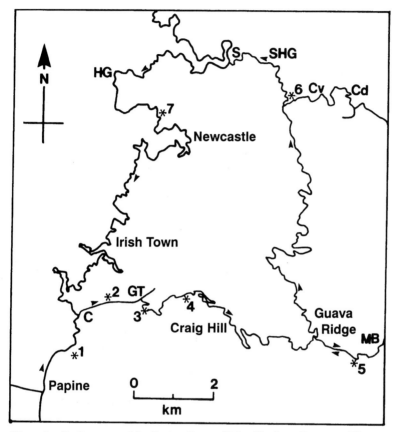

Figure 8: Outline map of the Papine-Guava Ridge-Hardwar Gap area. The main Papine-Newcastle road (B1) is shown as a heavy line; other roads are indicated by slightly finer lines.
Key: *=locality (1-7 refer to stops discussed in text); C=Cooperage; Cd=Clydesdale; Cv=Chestervale; GT=Gordon Town; HG=Hardwar Gap; MB=Mavis Bank; S=Section; SHG=Silver Hill Gap.
Arrows indicate the recommended route of the excursion.

Eastern and Central Jamaica

The sedimentary rocks of the Wagwater Group are predominantly siliciclastics, formed mainly of the purple to red conglomerates and sandstones of the Wagwater Formation, and the 'flysch'-like Richmond Formation (Robinson *et al.*, 1970) (Figure 1). Wescott & Ethridge (1983) recognised six major facies in the Wagwater and Richmond Formations, which they interpreted to have formed in a fan delta-submarine fan environment. Associated with the sedimentary rocks, and within the middle subdivision of the Wagwater Group (Green, 1977), are lava flows and pillow lavas of the Newcastle and Halberstadt Volcanic Formations, respectively. The Newcastle Volcanic Formation has been classified as a sequence of keratophyres and dacites, while the Halberstadt Volcanic Formation comprises spilites and basalts (Roobol, 1972; Jackson & Smith, 1979).

Stop 1. Kintyre Porphyry and Wagwater Fault Zone

Drive north from the University of West Indies campus at Mona and through the village of Papine onto the Gordon Town Road. Stop at the bottom of Skyline Drive. The Kintyre Porphyry is exposed in the bed of the Hope River (NGR 779524). This unit is a leucocratic, porphyritic-aphanitic (fine grained groundmass) rock that contains phenocrysts of plagioclase, hornblende and minor quartz. The porphyry intrudes the Wagwater Formation and has a chemistry that is indistinguishable from lavas of the Newcastle Volcanic Formation. The exposure is strongly jointed, and is extremely sheared and crushed along its western margin, which marks the boundary of the Wagwater Fault Zone. Disseminated pyrite, along with veins of gypsum, calcite and sulphur, occurs within the crushed zone, which is approximately 75 m wide (Carby, 1985). In this zone the minerals in the volcanic rock are altered, with calcite replacing plagioclase, and chlorite and haematite replacing hornblende. The cuprophilic fern *Pityrogramma calomelanos* (see Burns, 1960) grows in the fault zone.

The Wagwater Fault was first activated during the early Cenozoic as a normal fault with downthrow to the northeast (Horsfield, 1974). Normal faulting occurred intermittently throughout the Palaeogene as a result of rifting and crustal extension within the Wagwater Trough. A shift from transtensional to transpressional tectonic activity during the early Miocene, created by strike-slip movement along the Plantain Garden-Enriquillo Fault, produced a reversed movement along the Wagwater Fault leading to the uplift of Jamaica and the structural inversion of the Wagwater Belt (Mann *et al.*, 1985; Mann & Burke, 1990).

Eastern and Central Jamaica

Stop 2. Wagwater Formation

From stop 1 travel north to Cooperage and then east towards Gordon Town. The Wagwater Formation is exposed as a series of beds dipping steeply to the east along the roadside near the village of Cooperage and along Wiltshire Road (NGR 786536). These 'red beds' are formed mainly of conglomerates and breccias with minor sandstones and mudstones. The roadside exposures show poorly-sorted conglomerates and breccias, with clasts ranging from cobble to boulder size. The conglomerates and breccias are clast-supported, that is, the fragments are in contact with each other. Volcanic clasts predominate, some of which were mineralised prior to deposition. At Wiltshire Road a limestone bed about 30 cm thick is contained within the sequence; conglomeratic beds at this exposure are well-sorted. Minor normal faults can be seen at this stop, with about 1 m displacement of the limestone bed.

Based on the classification of Wescott & Ethridge (1983), the rocks along the main road section at Cooperage belong to their facies II (lower fan delta environment) and comprise conglomerates (approximately 75%), sandstones (5-15%) and shales (up to 20%). The presence of the limestone, associated with better-sorted conglomerates, suggests that there is a transition to facies III, which is characterised by thin limestone beds containing a shallow-water fauna, and it is interpreted as the coastal transition zone from alluvial fan to marine shelf deposits.

Stop 3. Richmond Formation

Drive to Gordon Town and turn right at the police station onto Craig Hill Road. Drive for 0.5 km along the road overlooking Gordon Town (NGR 797535). Stop 3 is an exposure of the Richmond Formation consisting of conglomeratic sandstones, sandstones and shales dipping to the west-southwest. The base of the outcrop contains thick beds of conglomeratic sandstones showing graded bedded features typical of Bouma turbidite divisions T_{a-c}, as well as scour marks and mudstone rip-up clasts. The conglomeratic sandstones contain subangular to rounded volcanic clasts. Reworked oysters are rare. This outcrop is assigned to facies V (proximal submarine fan).

Stop 4. Gypsum deposit and Newcastle Volcanic Formation

Drive along Craig Hill Road for approximately 1.5 km. Almost vertically-dipping strata of the Wagwater and Richmond Formations outcrop at this locality (NGR 809538), interbedded with a 7 m thick gypsum horizon. The gypsum has a grey colour, is laminated and is one of several similar beds

Eastern and Central Jamaica

that outcrop within the south-central Wagwater Belt. Commerial quantities of gypsum are located about 4 km to the south, where the thickness has increased to 60 m. Holliday (1971) considered that these evaporites were coastal sabkha accumulations, whereas Allen and Neita (1987) regarded them as continental sabkha (playa) deposits.

50 m to the east of this exposure the Newcastle Volcanic Formation is seen to overlie the Wagwater Formation. The volcanics are leucocratic, have a porphyritic-aphanitic texture, and contain phenocrysts of hornblende, plagioclase and quartz in a fine-grained groundmass. A zone of baking occurs below the contact between the two rock types and epidote, sulphur and chlorite occur as veins. According to Green (1974, 1977), this exposure represents the lowest of three major, silicic lava flows. He recognised at least three major flows (up to 600 m thick), together with several minor flows (150 m thick or less), in the Newcastle Volcanic Formation. These flows are thickest and most concentrated within the eastern half of the south-central portion of the belt (Jackson & Smith, 1979), where at least three volcanic centres have been recognised (Roobol, 1972).

Stop 5. Mineralised Halberstadt Limestone

Continue to drive east to Guava Ridge. At the Guava Ridge intersection take the road to Mavis Bank.

A wide variety of epithermal hydrothermal mineralisations occur within the Wagwater Belt (Kesler *et al.*, 1990), which are most commonly seen at and near the contact between igneous and sedimentary rocks (Fenton, 1981). 2 km from the Guava Ridge intersection (NGR 851523) there is a faulted contact between the Newcastle Volcanic Formation and the shallow water Halberstadt Limestone (Holliday, 1971; Green, 1977). The limestone here is about 10 m thick, dipping at about 60°E and steepening to almost 90° at the fault due to drag movement. Disseminated and veined sulphide minerals, such as pyrite and chalcopyrite, together with haematite, azurite and malachite, occur within the brecciated zone of the Newcastle Volcanic Formation.

Vein deposits of mainly magnetite are exposed along a track just east of the faulted contact and may be a continuation of the roadside outcrop. The vein is about 10 m wide and occurs at the volcanic-limestone contact.

Stop 6. Yallahs Fault Zone

Return to the intersection at Guava Ridge and turn right to St. Peters. Stop at the road junction to Chestervale and Clydesdale (NGR 833586).

Eastern and Central Jamaica

The Yallahs and Wagwater Faults have similar structural histories, with normal faulting in the Palaeogene followed by reverse faulting in the Neogene (Mann et al., 1985; Mann & Burke, 1990). At St. Peters, mafic volcanic rocks have undergone cataclasis, caused by tectonic movement along the Yallahs Fault Zone. These volcanic rocks have been previously described as the Silver Hill Schists (Matley, 1951) and the Green Volcanics (Green, 1977), and were considered to be Cretaceous in age (='St Peters Inlier'; Figure 2). However, geochemical data suggest that they are the Palaeogene Halberstadt Volcanic Formation (Jackson, 1986). Where the original texture has been preserved, the rock is fine grained with plagioclase laths and granular olivine. Post-magmatic alteration is evident throughout most of the outcrop with the presence of calcite, epidote and quartz veins, and the devitrification and chloritization of the groundmass to form spilites.

The volcano-tectonic setting of these rocks indicates an intraplate environment. Geochemical analyses of the Halberstadt Volcanic Formation prompted Jackson & Smith (1979) to conclude that these rocks were comparable to plateau-type basalts because of their similar concentration of light rare-earth elements (LREE) and large ion lithophile elements (LIL). However, more recently, Stern et al. (1990) have shown that back-arc basalts of the Mariana Trench in the western Pacific Ocean, which are analogous to the Halberstadt Volcanic Formation, had their REE and LIL enrichment signatures produced either by the mixing of older arc-related mantle and mid-ocean-ridge basalt (MORB) sources or by metasomatism of a MORB-like source, during the early evolutionary stage of crustal extension. The Halberstadt Volcanic Formation is therefore now classified as enriched back-arc basin (BAB) basalt that evolved within a narrow rifted area.

Stop 7. Metamorphosed Newcastle Volcanic Formation

Drive to Silver Hill Gap, then turn left and proceed to Section. Turn left at Section and proceed to Hardwar Gap, where massive lava flows of the Newcastle Volcanic Formation are exposed at the roadside (NGR 801583). Most of the section from Hardwar Gap to Newcastle consists of quartz keratophyre lava flows (Jackson, 1977). The rocks display a volcanic texture, but contain a low grade metamorphic mineral assemblage. The rocks contain 'porphyroclasts' of albite, chlorite and minor quartz. The groundmass mineralogy includes granules of quartz with laths and microlites of albite, chlorite and sericite. Texturally these rocks are no different to the calc-alkaline dacites of the Newcastle Volcanic Formation, but they differ in their mineralogy and, to a lesser extent, geochemistry. The most pronounced chemical differences are in the proportions of Na_2O, K_2O, CaO, Sr and Ba. The mineralogical and chemical dissimilarities are

Eastern and Central Jamaica

attributed to post-magmatic metasomatism of the dacites which took place under conditions of high P_{CO_2}. Low shear stress and temperatures below 400°C were responsible for the retention of the volcanic textures (Jackson & Smith, 1978).

Return to Kingston from Newcastle via Irish Town. Along the scenic route there are more outcrops of the Wagwater, Richmond and Newcastle Volcanic Formations. Hydrothermal alteration is also evident along this section, with veins of barytes occurring within brecciated sections of the Newcastle Volcanic Formation near Irish Town, together with hydrothermally altered clay minerals (Scott & Drakapoulos, 1989).

NORTHEAST COAST

Drive north from Kingston on the main A3 road (=Junction Road) and continue east to Buff Bay on the north coast (Figure 9).

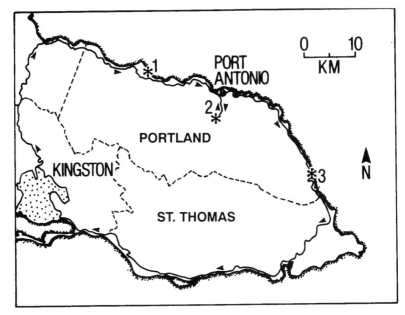

Figure 9: Outline map of eastern Jamaica, showing Junction Road (left) and the coast road (after Donovan et al., 1990, fig. 1A). Arrows mark the route taken by the field excursion to northeast Jamaica. Stops 1 to 3 are indicated. The two eastern parishes are Portland (north) and St. Thomas.

Eastern and Central Jamaica

Between Buff Bay and Hope Bay, the road turns inland. A left turn is taken at Black Hill, towards the coast. The geology of the Low Layton volcanics may be examined by walking along the closed railway trackbed on the coast (Donovan *et al.*, 1990; NGR 937741 and to the southeast).

Stop 1. Low Layton

The Low Layton volcanics form part of an east-west ridge about 2.5 km in length which intersects the north coast of Jamaica at Hope Bay. The Low Layton volcanics are the only evidence of post-Palaeogene volcanism on the island, comprising a sequence of alternating pillow lavas, pillow breccias and hyaloclastites which appear to have solidified in about 100 m of water (Roobol, 1972). These rocks have a basaltic composition (Roobol, 1972; Roobol & Horsfield, 1976; Jackson & Smith, 1982) and, on the basis of their major and trace element chemistry, belong to the alkali magma series. The texture of the rock is generally aphyric and vesicular. Wadge (1982) noted that ". . . a typical hand specimen of lava has 10-30% vesicles arranged in bands, largely filled with calcite, and brown microphenocrysts of iddingsitised olivine". The breccias are interbedded with the lavas and contain angular fragments of basalt in a yellow-brown sideromelane/palagonite (partly altered basaltic glass) matrix (Wadge, 1982).

The Low Layton volcanic rocks outcrop in an area of no more than 3 km^2, with the total volume of volcanic rocks above sea-level being approximately 9×10^7 m^3 (Wadge, 1982). These volcanic rocks occur within a succession of foraminiferal limestones of the Lower Coastal Group (Robinson, 1958b), but contain disrupted chalky limestones of the mid-Tertiary Montpelier Group within the sequence of lava flows and breccias. Robinson (1958b, p. 70; 1959) considered the foraminiferan fauna of the overlying sedimentary rocks to be similar to that of the top of the Buff Bay Formation, suggesting the presence of the upper N16 to lower N17 biozones (Banner & Blow, 1965; Robinson, 1967a, pp. 34, 36; 1969, pp. 3-5), that is, low in the Upper Miocene (Harland *et al.*, 1990, fig. 3.15). This is in good agreement with a K-Ar isotopic date of 9.5 ± 0.5 Ma (Wadge, 1982) for the volcanic rocks.

The eruption of the Low Layton volcanics coincided with major changes in sea-floor spreading that were taking place in the Cayman Trench to the north of the island causing extrusion along the eastward extension of the Swan Transform Fault, now referred to as the Plantain Garden-Enriquillo Fracture Zone.

Eastern and Central Jamaica

Stop 2. Fellowship

Return to the main A4 road and drive east to Port Antonio. Take the minor road south through Breastworks to Fellowship (Figures 9 and 10). Just after Fellowship a turning onto a track on the right leads to stop 2 (NGR 071651), where the Lower Palaeocene (NP1; E. Robinson, personal communication) Richmond Formation (='Moore Town Shales' of Jiang & Robinson, 1987) is exposed in steep cliff faces immediately upstream of the bridge over the Rio Grande. Here, the Richmond Formation is turbiditic and is considered to have been deep-water in origin. Mudstones are blocky, dark and often carbonaceous, interbedded with fine- to medium-grained sandstones. The mudstones are generally in thin layers (less than 2 cm thick) and are either finely-laminated or massive, with slight changes in colour. Sandstone interbeds are up to 30 cm thick (but typically less) and are laterally continuous. Internally they exhibit partial Bouma sequences, including common Bouma T_b, less common T_{b-c} and rare T_{a-b} or T_{a-c} (Donovan *et al.*, 1990).

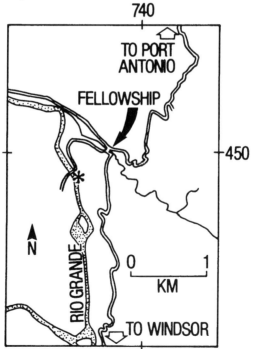

Figure 10: Outline map of the Fellowship area, parish of Portland, showing the position of stop 2 (after Donovan et al., 1990, fig. 3).

Eastern and Central Jamaica

Basal surfaces of sandstones are planar and may be erosive, as indicated by rare flute and groove structures, though are more typically non-erosive. Some of the sandstone layers include comminuted bioclastic debris on their soles (Donovan & Veltkamp, 1992). The contact with the underlying Upper Cretaceous is just north of this locality, but is faulted, unfortunately. Indeed, the Cretaceous/Tertiary boundary has yet to be detected in Jamaica, in contrast to good sections which have been described from Haiti and Cuba.

Trace fossils are relatively abundant on many of the soles, and some upper surfaces, of the sandstone beds (locality 5 of Pickerill & Donovan, 1991). The ichnotaxa identified from this locality include abundant *Palaeophycus tubularis* Hall, *Planolites beverleyensis* (Billings) and *Trichichnus linearis* Frey; rare to common *Helminthopsis* isp. and *Scolicia* cf. *plana* Książkiewicz; and rare *Palaeophycus striatus* Hall and *Phycodes* isp. These ichnotaxa suggest that the depositional site was occupied by predaceous or suspension-feeding organisms, probably annelid worms, which produced *Palaeophycus*, and mobile deposit feeders, most likely polychaete annelids or priapulids, responsible for *Planolites* and *Helminthopsis*. The presence of *Scolicia* suggests the former presence of spatangoids (heart urchins), though no irregular echinoid macrofossils are known from the Richmond Formation (Donovan & Veltkamp, 1992). This assemblage is typical of the *Scolicia* ichnocoenosis (trace fossil assemblage) of Pickerill and Donovan (1991), which is associated with deeper water palaeoenvironments.

Stop 3. Manchioneal

Return to the A4 road and continue east, then southeast, to Manchioneal (Figure 9). Stop 3 is a line of old sea cliffs adjacent to the road (NGR 262536). This sequence is the type section of the Manchioneal Formation of the Coastal Group, originally described by Trechmann (1930). It is formed of nodular limestones interpreted as open marine in origin, analogous to the carbonate slopes in the northern Bahamas (Mullins *et al.*, 1980) and broadly similar in sedimentological style to unit A of the Round Hill Beds (see p.10). The fauna at this locality is unusual (for Jamaica) in including a moderate diversity of brachiopods (Harper & Donovan, 1990), which may have been attaching to nodules following erosion of overlying soft sediment. Other elements of the fauna include benthic foraminifers, solitary and colonial corals, echinoids and benthic molluscs. Robinson (1967a, 1969) considered this formation to be early Pleistocene in age.

Eastern and Central Jamaica

SOUTHEAST COAST

Drive east from Kingston on the main A4 south coast road and take the turning towards Easington, which is before milepost 17 (Figure 11).

Stop 1. Yallahs fan delta

A roadside exposure about 2.5 km south of Easington Bridge (about NGR 929385) gives a commanding view of the Yallahs fan delta, although much of the detail is obscured by vegetation. However, the outcrop at this site appears to be a remnant of a previously more extensive fan and Robinson (1965) included this exposure within the Pleistocene Upper Coastal Group. The lithologies present are identical to those of the more recent fan having been derived from the same sedimentary, metamorphic and igneous rocks of the Cretaceous Blue Mountain Inlier and the Palaeogene Wagwater Group (the latter including an Eocene fan delta sequence).

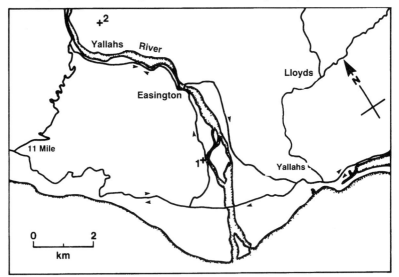

Figure 11: Outline map of the Yallahs district, parish of St. Thomas.
Key: +=locality; 1=overview of Yallahs fan delta; 2=Judgement Cliff landslide (view from road); heavy lines=principal roads; stippled lines=coast, salt pond (to southeast) and Yallahs River. The arrows indicate the route travelling east (via localities 1 and 2) and returning to Kingston along the main south coast road (A4).

Eastern and Central Jamaica

The lobate Yallahs fan delta has an area of about 10.5 km². The fanhead is at Easington, 2 km from the coast. The Yallahs River, which supplies the sediment for the fan delta, rises at a height of 1000 m in the southern Blue Mountains and drains a high relief, sediment-rich hinterland which experiences high to very high seasonal rainfall (Wescott & Ethridge, 1980). Maximum sediment thickness (30 m) occurs towards the fanhead.

The submarine part of the fan delta extends to a depth of 1100 m below sea-level over a horizontal distance of 4000 m. The fan delta forms the northern margin of the fault-bounded Yallahs submarine basin, a basin now being filled by non-carbonate material from both the Hope-Liguanea and Yallahs fan deltas (Burke, 1967; Wescott & Ethridge, 1980).

The abundance of large boulders at the Yallahs fording (which is crossed on the way back to Kingston) gives some indication of the strength of flood currents. It is estimated that hurricane discharges can transport 2330 m³ s⁻¹, compared with the normal rainy season discharge which ranges from 0.00037 to 17 m³ s⁻¹. Catastrophic events dominate the physical processes of tropical streams. This is in contrast to temperate streams, where a constant, high-level background flow reworks flood-event sediments fairly rapidly.

The main sedimentary units are low-relief longitudinal gravel bars. During non-flood periods flow is restricted to a few shallow streams between the bars and may not reach the shore due to percolation and removal by man. Relief is essentially that of the largest clast. Structures in the sandy gravels of the river bed are crude horizontal bedding and imbrication. Flood plain deposits are grey, very fine grained, cross-laminated sands.

Stop 2. Judgement Cliff landslide

Drive north, past Easington Bridge, and continue along the Yallahs River valley. The Judgement Cliff landslide is best viewed from the football pitch at about NGR 915434 (Figure 11).

During tropical storms and hurricanes, high precipitation initiates landslides in a variety of rock types and geomorphic settings throughout Jamaica (Maharaj, 1990; Manning *et al.*, 1992). That catastrophic landslides have occurred on the island during the prehistoric and historic past is attested by the occurrence of ancient slip masses. For example, the Judgement Cliff landslide occurred during 1692, supposedly after the major earthquake (June 7th) of that year and probably during a hurricane (October 18th-19th), which was accompanied by exceptionally heavy rain (Zans, 1959). The slide bulge is formed of limestones of the Eocene Bonny

Eastern and Central Jamaica

Gate Formation, White Limestone Supergroup (Robinson, 1967b), which have slipped over a basal unit of mudrocks and evaporites which in turn overlie the sandstone-shale sequence of the Richmond Formation (Zans, 1959). The eastern boundary of the slide is marked by a high-angle fault. The estimated mass of the landslide (recalculated after Zans, 1959) is 165×10^9 kgm.

Stop 3. The Port Morant Formation - type section

Return to and cross Easington Bridge, rejoining the main A4 road at Yallahs. Continue east and note the hypersaline lagoons of the two Great Salt Ponds immediately outside the town. Drive east to Port Morant.

Exposed along the road and seacliffs just west of Port Morant (NGR 216377) are Port Morant Formation pebbly sandstones showing good evidence of channeling that are interpreted as alluvial/fluvial sediments. The beds are thought to be late Pleistocene in age (last interglacial, equivalent to the Ipswichian of the British sequence).

Stop 4. The Bowden Shell Bed

Continue east to the Bowden turnoff, marked by a triangular lawn with railings. Drive southwest on the Bowden road. The Bowden Shell Bed is exposed in the bank on the left just before the junction of the track to Old Pera (NGR 225377).

This Pliocene horizon is the most famous fossiliferous deposit in Jamaica, containing a fauna of particularly well-preserved shelly fossils, particularly benthic molluscs. Woodring (1925, 1928) identified about 600 species of bivalves, gastropods and scaphopods from this horizon. On the basis of the mollusc fauna, this bed was previously considered to be Miocene or even Oligocene in age. More recently, foraminiferal studies by Robinson (1969) and others have indicated that the shell bed is actually Upper Pliocene (zone 20 of Blow, 1969). Apart from the large diversity of molluscs, the fauna also includes foraminifers (Palmer, 1945), bryozoans (Lagaaij, 1959), calcareous algae (Rácz, 1971), echinoids (S.K.D., research in progress) and corals. A single fragmented needlefish jawbone has been described from the same horizon by Caldwell (1969).

Robinson (1969) interpreted the environment of the shell bed as being of relatively deep water (greater than 100 m) based on the planktonic foraminifers. These deep waters were fed by detrital material of shallow water or terrestrial origin by sliding and sand flow mechanisms, an inference supported by the occurrence of rounded pebbles, fragments of wood, shallow water, massive hermatypic corals and freshwater molluscs.

Eastern and Central Jamaica

Stop 5. The Old Pera beds and Port Morant Formation

The track from Bowden to Old Pera is only suitable for vehicles with 4-wheel drive. Otherwise return to the A4 road and continue east. At Phillipsfield turn southeast to New Pera, and then southwest and west through the cane fields to Old Pera. Park at the junction with the track from Bowden (NGR 222359) and continue west to the coast on foot (Figure 12).

Exposed along coastal cliffs north and (particularly accessible) south of Pera Point, the Port Morant Formation overlies the Old Pera Beds. Robinson (1969) correlated the Old Pera beds with the Manchioneal Formation (see p.26) of early Pleistocene age. The Old Pera beds are a sequence of storm-influenced, siliciclastic, sedimentary rocks, with prominent sandstone beds. The fauna is varied, including the trace fossil *Bichordites monastiriensis* (Pickerill *et al.*, 1993c), benthic molluscs, sand dollars and other echinoids (Donovan *et al.*, 1994a), barnacles and corals.

The base of the Port Morant Formation is marked by a boulder conglomerate. All of the boulder-sized clasts are limestone mainly derived from land sources (terrigenous). The base appears to truncate the underlying Old Pera beds. Immediately overlying the boulder conglomerate there is a rapid facies change into heavily bioturbated, brown siliciclastics with some marlier horizons. The trace fossils *Thalassinoides, Ophiomorpha* and *Palaeophychus* are common. Further up-section there is an influx of terrigenous pebbles which form stringers and two thicker (0.5 m) conglomerate horizons. The top of the Port Morant Formation is capped by a highly fossiliferous 'raised beach' deposit which have been correlated with the Falmouth Formation on the north coast (a last interglacial raised reef; see above).

The lack of primary sedimentary structures in this section makes interpretation difficult. However, the Port Morant Formation at this locality may represent the infilling of a back-barrier/lagoon. The base is marked by large platy slabs which appear to be almost *in situ*, which may indicate a subaerial weathering surface. The abundance of large boulders indicates a high energy environment and it is suggested that this may have formed part of a back-reef flat or pavement. Large corals, both *in situ* and overturned, are buried by the overlying succession. The absence of a normal marine fauna, high level of bioturbation and fine grain size all suggest lagoonal sediments. The horizons of terrigenous pebbles probably represents fluvial influxes which may have been reworked to form pebbly beaches. The sequence seen at stop 3 indicates the presence of a nearby alluvial/fluvial system which could have supplied this material.

Eastern and Central Jamaica

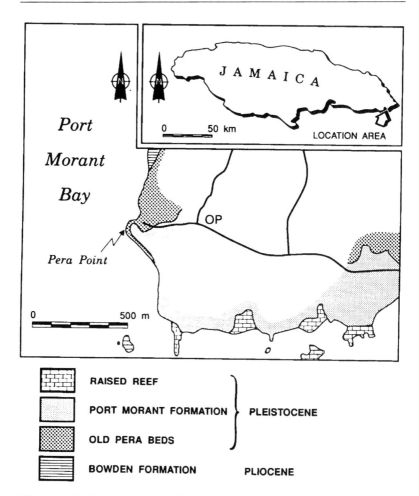

Figure 12: Geological map of the Old Pera district, parish of St. Thomas (slightly modified after Donovan et al., 1994a). OP=Old Pera (park at the T-junction). The inset map shoes the position of Old Pera within Jamaica.

Eastern and Central Jamaica

BAUXITE MINES

The Alcan Jamaica Company are happy to arrange for geological parties to tour their bauxite mining operation at Kirkvine, parish of Manchester. Potential visitors should first contact The Corporate Relations Department, Alcan Jamaica Company, Kirkvine, P.O., Mandeville, parish of Manchester, Jamaica. Bauxite was only discovered to be present in commercial quantities in Jamaica in 1942. Production started in 1952, since when the island has become a leading exporter of bauxite and alumina.

Exploitable bauxites are found mainly in central Jamaica on the Clarendon Block (Figure 3). These bauxites are of the terra rossa type and occur exclusively as pocket or blanket deposits on the karstified surface of the mid-Tertiary White Limestone Supergroup (Robinson, 1971). The contact between bauxite and the underlying limestone is sharp. The bauxite is generally red, with localised occurrences of off-white and yellow bauxites overlying some white limestones (Anderson, 1971). The thickness of these deposits varies between 1 and 50 m (Porter et al., 1982). The best grade of bauxite ore is located at elevations above 300 m, and where there is a significant separation between the deposit and the water table (Hill, 1955).

Jamaican bauxites contain four economically important minerals: gibbsite, $Al(OH)_3$; boehmite, $AlO(OH)$; aluminium goethite, $(Fe, Al)O.OH$; and haematite, Fe_2O_3. The main chemical constituents of the high grade ore are summarised in Table 1. In Jamaica, high grade ore is often blended with bauxite of a slightly lower quality that contains silica values above 4%.

TABLE 1: Principal chemical constituents of Jamaican bauxite.

alumina	46-52%
combined water	24-28%
iron III oxide	17-22%
titania	2-3%
silica	<4%
phosphorus pentoxide	<3%
lime	<1%
manganese oxide	<1%

Eastern and Central Jamaica

The origin of Jamaican bauxite has been a point of discussion for over 40 years. There are three principal hypotheses: the residual theory, the alluvial theory, and the volcanic ash theory. The residual theory proposes that insoluble residues have remained after the dissolution of the limestone by tropical weathering. The formation of a residue coincided with the uplift of Jamaica approximately 10 million years ago. It is estimated that more than 250 m of limestone would have to be weathered in order to supply the necessary residue for bauxitization (Sinclair, 1966; Robinson, 1971).

The alluvial theory suggests that the source of the bauxite was from the older Cretaceous inliers, formed mainly of igneous and igneous-related rocks (Zans, 1958). Aluminium-rich minerals from these inliers were transported via subterranean streams in the younger limestones, in which they were eventually deposited. Subsequent exposure was a consequence of uplift and erosion of the limestones.

Comer (1972) formulated the volcanic ash theory. He postulated that the source of the bauxite was from the weathering of volcanic ash that was deposited on karstified limestone during the Neogene. The presence of bentonite beds within Upper Miocene limestones (see above) indicates that a substantial amount of air-fall ash was deposited below sea-level and, by inference, also above it at that time.

ST. MARY COAST

Drive north from Kingston on the main A3 road (=Junction Road). Before Annotto Bay, the road turns northwest. Drive through Port Maria. Stop 1 is an old sea cliff inland of the road, beneath Fort Baldano at Point Haldane (NGR 615917; Figure 13).

Stop 1. Port Maria Shell Bed

The Port Maria Shell Bed forms part of the Lower Eocene Port Maria Member (Mann & Burke, 1990), Richmond Formation, Wagwater Group. Trechmann (1924, p. 8) published a measured section that included this horizon. The Richmond Formation *sensu lato* ranges from the Palaeocene to the Lower Eocene (Robinson & Jiang, 1990; Pickerill & Donovan, 1991), includes a variety of sedimentary facies indicative of shallow to deeper water environments, as indicated by the sedimentological (see 'Southern Wagwater Belt', p.17) and ichnological evidence, and was deposited within a large submarine fan system.

Eastern and Central Jamaica

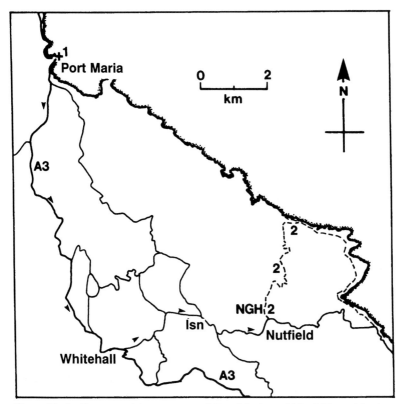

Figure 13: Outline map of the area around Port Maria and Nutfield, parish of St. Mary. Key: +1=locality 1, Port Maria shell bed; 2-2-2=traverse from Nutfield to coast; Isn=Islington; NGH=Nutfield Great House. Coast stippled; main A3 road heavy line; minor roads light lines (not all shown); track dashed. Arrows show the recommended route from Port Maria to Nutfield.

Macrofossils are generally poorly represented in strata of the Richmond Formation, although trace fossils are locally common and moderately diverse (Pickerill & Donovan, 1991; Pickerill *et al.*, 1992, 1993b). Recent discoveries of body fossils from the Port Maria area include a regular echinoid (Donovan & Pickerill, 1993) and an articulate brachiopod (Donovan *et al.*, 1993b). However, the most fossiliferous horizon known from the formation is the Port Maria Shell Bed. This bed is about 9 m thick, and is a siliciclastic, conglomeratic to sandy, shelly unit best known

Eastern and Central Jamaica

for its transported fauna of Eocene benthic molluscs (Trechmann, 1924) and reworked Upper Cretaceous rudist bivalves (Chubb, 1971). H.L.D. recently collected only the second Jamaican Eocene crinoid from this locality (Donovan et al., 1994b). Uplift since the late Pleistocene is indicated by the raised reef (=Falmouth Formation?) at the cliff top.

Stop 2. Nutfield to Stileman's Cove

Return to Port Maria on the A3 road and continue south. Near Whitehall, turn northeast on the road to Islington and Nutfield. Take the branch road to Nutfield Great House (about NGR 675841; Figure 13).

This part of the field trip involves a traverse from Nutfield to the north coast (about 3 km) and then along the coast for about 1 km. The volcanic rocks in the vicinity of Nutfield outcrop in the northern part of the Wagwater Belt (Figure 3), and have a similar age, mineralogy and chemistry to the volcanic rocks of the Newcastle and Halberstadt Volcanic Formations found further to the south. The section at Nutfield shows a pillow lava flow of mafic composition, overlain by a dacite lava flow. The entire unit, which is approximately 100 m thick, is contained within the Lower Eocene Roadside Member of the Richmond Formation (Mann & Burke, 1990).

The pillow lavas have a mineralogy and chemistry that is characteristic of spilitic rocks and not unlike the spilites of the Halberstadt Volcanic Formation. These rocks possess high Na_2O/K_2O ratios (Table 2), with 'porphyroclasts' of chlorite and albite in a fine-grained groundmass of chlorite. At some of the outcrops around Nutfield, the pillows display radial jointing, and contain amygdales of chalcedony, calcite and chlorite. Based on the average vesicle size, Roobol (1972) concluded that these lavas, which were originally basaltic (Jackson, 1977), were erupted in about 300 m of water.

The silicic lava flow that outcrops in the area of Nutfield is petrologically the same as the dacites of the Newcastle Volcanic Formation. It displays a porphyritic-aphanitic texture in which there are phenocrysts of plagioclase feldspar, hornblende and clinopyroxene in a groundmass of plagioclase microlites and chloritised glass. Chalcedony occurs as a secondary mineral infilling vesicles and fractures within the flow. Along the road from the Nutfield Great House to the coast there is an exposure of the dacite lava flow, about 3 m thick, which displays columnar jointing. The flow is overlain by volcanic breccia containing clasts similar in composition to the flow (Table 2).

Eastern and Central Jamaica

TABLE 2: Major element analyses of the volcanic rocks of Nutfield, parish of St. Mary (Jackson, 1977).

	JA76	JA78	JA75	JA80
SiO_2	48.53	48.31	61.41	65.79
Al_2O_3	17.94	13.24	15.45	14.61
FeO*	7.94	6.39	4.10	2.97
MgO	6.07	4.29	1.82	2.01
CaO	6.11	11.25	5.27	4.39
Na_2O	3.51	3.79	2.95	3.91
K_2O	0.86	0.39	0.89	3.83
TiO_2	1.07	1.51	0.71	0.47
P_2O_5	0.16	0.10	0.25	0.16
MnO	0.08	0.10	0.08	0.14
H_2O**	5.16	3.08	6.21	3.04
CO_2	3.29	6.88	0.26	1.16

*=total iron oxide
**=total water

JA76=spilite pillow lava, Nutfield.
JA78=spilite clast in breccia, Stileman's Cove.
JA75=dacite lava flow, Nutfield.
JA80=dacite clast in breccia, Stony Cove.

In association with the flows are a series of volcanic breccias, most of which occur within the Richmond Formation. These breccias are best exposed along the north coast. They are interpeted as collapse breccia deposits that formed on the flanks of the volcanic pile whose centre was located near Nutfield (Roobol, 1972; Mann & Burke, 1990).

CENTRAL INLIER

Drive west from Kingston on the main A2 road. Take the third (and final) exit to May Pen. Continue north to Chapelton on the B3 road.

Stop 1. Chapelton Formation

The Chapelton Formation, Yellow Limestone Group, which ranges from the mid-Lower to mid-Middle Eocene, is formed of at least five members

Eastern and Central Jamaica

deposited under fluvial and marine conditions (Robinson, 1988). The yellow colour is produced from impurities derived from weathered Cretaceous rocks. After a period of general emergence in the Palaeocene, the island had started to become inundated again in the Eocene until it was completely submerged. This total submergence marked the commencement of deposition of the White Limestone Supergroup, which generally consists of very pure limestones. The Chapelton Formation is well-known for its large and diverse fossil biota, particularly in the marine limestones. The White Limestone Supergroup often appears to have a comparatively impoverished macrobiota, but this may be partly due to its high level of induration making fossils less obvious and less easy to collect.

At this locality, in the playing field of Clarendon College (NGR 212592), a contact is exposed between the Chapelton Formation and the overlying Troy Formation of the White Limestone Supergroup. The Chapelton Formation is quite fossiliferous, and yields a fauna including bivalves, corals, foraminifers and rare echinoids. The apparent curved structures crossing the contact may be joints.

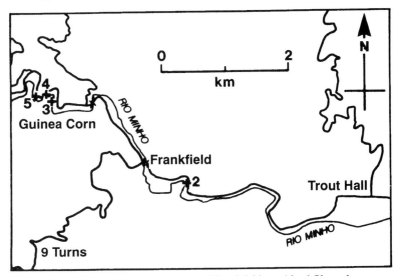

Figure 14: Outline map of the area around Frankfield, parish of Clarendon. Key: +=locality (2-5 refers to stops discussed in text); heavy lines=main roads; finer lines=Rio Minho (note that there are numerous streams that feed this river, but which are not shown).

Eastern and Central Jamaica

Continue on the B3 towards Frankfield. Stop 2 (approximate NGR 118656; Figure 14), an exposure below road level and in the bank of the Rio Minho, is between Trout Hall and Frankfield.

Stop 2. Summerfield and Guinea Corn Formations

The Summerfield Formation is a sequence of unfossiliferous, siliciclastic sedimentary rocks (with rare, devitrified ignimbrites; Roobol, 1976), composed almost entirely of hornblende andesite fragments. The base of the Summerfield Formation is Maastrichtian (late Cretaceous), as it sits with apparent conformity on the underlying Guinea Corn Formation of late Campanian to early Maastrichtian age (Robinson, 1994). However, an ignimbrite from high in this unit has been dated as Palaeocene (55.3±2.8 Ma) based on fission-track age determinations (Ahmad et al., 1987a).

The lowermost part of the formation, including its contact with the Guinea Corn Formation, is exposed at this locality. It is approximately 100 m thick, consisting of fairly regularly bedded, pinkish sandstones and gritstones. Roobol (1976) described the beds as being crystal-rich, comprising 20-30% labradorite (An_{68}), 2-10% brown amphibole, 1-3% iron-titanium oxide and 1% embayed quartz. Many of the larger megacrysts are broken. The megacrysts are contained in a very fine-grained, devitrified matrix, lacking glass and shards. Lithic inclusions comprise 2-12% of the rock.

The overall thickness of the formation is around 460 m. The lower part, which is pumice-poor, interdigitates with marine fossiliferous sediments of the Guinea Corn Formation. The upper part (360 m) is pumice-rich and interpreted as being fluviatile. The Summerfield Formation is interpreted as representing a single episode of hornblende-andesite pyroclastic volcanism based on its monotonous mineralogy.

Stops 3-5. Guinea Corn Formation

The Guinea Corn Formation is a sequence of grey, rubbly limestones and carbonate mudrocks with interbedded shales, volcaniclastic rocks and some thin lignites. The shales are generally unfossiliferous, but the limestones have yielded a diverse fauna. This formation is best know for its diverse assemblage of rudist bivalves (Chubb, 1971), but it also includes other benthic bivalves, gastropods (see, for example, Sohl, 1992), foraminifers, ostracods, corals (Coates, 1977), echinoids (Donovan, 1993b), and rare barnacles and crabs.

Stops 3-5 are all in the river bed of the Rio Minho, adjacent to the B3 road west of Frankfield. Stop 3 (NGR 099670; Figure 14), about 1½ km

Eastern and Central Jamaica

west of Frankfield, is an exposure of steeply-dipping beds on the opposite side of the river to the road. This is the only locality in the Central Inlier where brachiopods have been collected. They have been attributed to *Dyscritothyris* (Harper & Donovan, 1990), a small (2-4 mm), smooth, pedunculate terebratulid. Only a dozen specimens are known, all coming from a single horizon of weathered shale. Monospecific, in-place horizons of the small, gregarious rudist *Radiolites* also occur together with numerous, apparently reworked, attached valves of the giant *Durania*.

Stop 4 (=stop 3 of Robinson, 1988; NGR 097670) contains some excellent rudist bivalves, as well as other shelly fossils. The rudists include the large, almost cylindrical *Durania*, the flat-lying and curved *Titanosarcolites* and smaller, conical *Radiolites*. This stop is essentially along strike from stop 3, and is exposed adjacent to the road on the opposite (northwest) side of a large horseshoe bend in the road and river.

The rudists developed many shell forms which can be divided between three adaptive groups (Skelton, 1985, 1991). Elevators were tall, conical to barrel-shaped shells in which the feeding margin was raised above the sediment surface. The small apical attachment site was usually buried in the sediment. Elevators grew either singly or gregariously in waters that were usually calm. The rapid upward growth (=elevation) of the margin may have been to raise the feeding region above a low fog of turbid water. The external appearance of an elevated rudist is often reminiscent of a solitary coral.

Clingers (Skelton, 1991), otherwise called encrusters (Skelton, 1985), were compact, usually bun-like forms with a broad, encrusting lower surface. Encrustation ranged from intimate adhesion on a hard substrate to moulded overgrowth on a stable sediment surface in current conditions that were sufficiently strong to transport mud and thus prevent burial. Unlike recumbents, stability was a direct function of the area of sediment contact.

Recumbents were large, horizontally-extended shells that lay freely on the sea floor and were very stable, even in mobile sediment. The extended shell enclosed a large 'virtual surface' which conferred gravitational stability. This stability was aided by a low centre of gravity.

Just upstream of the last locality is an exposure (stop 5) in which rudist-bearing beds with common *Titanosarcolites* are present. Individual beds are laterally continuous and it is possible to identify repeated sequences of three units A, B and C. Unit A is a carbonate-rich mudstone with few macrofossils. Unit B is a nodular, marly limestone with clusters of elevating radiolitids and abundant shell debris, including fragments of *Titanosarcolites*, which often serve as a substrate for the attachment of corals and other rudists. Unit C is a massive limestone with rudist clingers and elevators. *Titanosarcolites* is found in life position at the top of unit C. These sequences are repeated throughout the section and are interpreted as representing prograding banks on a shallow carbonate shelf (Skelton *et al.*, 1992). The cyclicity of the repeated sequences suggests that these organisms were in dynamic equilibrium with the environment.

Rudists have been interpreted as reef-forming organisms (Kauffman & Johnson, 1988) and the Guinea Corn Formation may be regarded as the 'type section' for rudist reefs (Kauffman & Sohl, 1974). However, Gili *et al.* (1990) have questioned this interpretation, and concluded that rudists were part of a mollusc-rich, carbonate shelf fauna and never built any reef structures. There is no evidence of any reef-like structure in the Guinea Corn Formation, nor is there any indication that these rudists have been transported from a reef to their present setting. The presence of lignites implies occasional hyposalinity and nutrient flux.

BROWN'S TOWN, ST. ANN

Drive west from Kingston on the main A1 road. This turns northwest at the roundabout after Spanish Town. Note the Bog Walk Gorge, an impressive limestone solution feature. Alcan's alumina plant is on the right before Ewarton. Stop 1 is on the right side of the road at the summit of Mount Diablo (NGR 399726).

Stop 1. Red mud lake

The red mud lake is a disposal site for waste resulting from the production of alumina (aluminium hydroxide) from bauxite by treatment with sodium hydroxide (caustic soda). Red mud is composed of iron oxide and excess caustic soda, and its elimination is a pressing environmental problem (Bell, 1986). The caustic soda enters the water table if not properly contained and has a large volume, making disposal difficult.

Stop 2. Lee's Marl Crushing Plant

Continue north to Moneague and take the left fork (A1) towards Claremont. At Green Park take the B11 to Bamboo and on to Brown's Town. Stop 2 is on the left about 1½ km before Brown's Town (NGR 125936; Figure 15).

The Tertiary limestones of Jamaica are biostratigraphically zoned principally on the basis of their included fauna of foraminifers (Hose & Versey, 1957). The various formations of the White Limestone Supergroup

Eastern and Central Jamaica

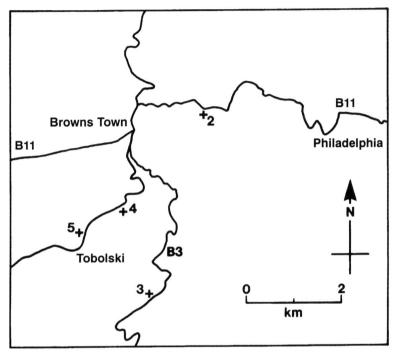

Figure 15: Outline map of the Browns Town area, parish of St. Ann, showing the principal roads. Key: +=locality; 2=Lee's Marl Crushing Plant; 3=Friendship; 4=disused bauxite pit, Tobolski road; 5=reclaimed bauxite pit near Tobolski.

(mid-Middle Eocene to low Upper Miocene), deposited in both shallow and deeper water palaeoenvironments, are purer and are usually more indurated than the various facies of the underlying Yellow Limestone Group. The Oligocene Brown's Town Formation is a shallow-water unit of the White Limestone succession and has a relatively wide areal extent. Wallace (1969) described the upper part of this formation as a ". . . foraminiferal-echinoid biomicrite . . . representing a transition from the fore-reef shoal environment into an open sea environment." The echinoid fraction is currently being studied by H.L.D. Previously, only one species of echinoid had been formally described from this formation, *Eupatagus hildae* Hawkins, 1927. Current research has so far recognised 12 species assigned to 5 different orders (Dixon & Donovan, 1992).

Mr. Lee's pit is scheduled to resume production in the near future and permission to enter should be obtained at his adjacent home. Quarrying has

Eastern and Central Jamaica

created a large face, about 100 m in length, of white, grey-weathering beds of sparry limestone interbedded with softer, biomicritic bands. Beds are laterally continuous, but somewhat friable. The predominant fossils are large benthic foraminifers, associated with echinoids, bivalves (particularly oysters, scallops and tubes of *Kuphus*), large asteroid ossicles, colonial scleractinian corals, sirenian ribs and bryozoans. The commonest echinoid is the sand dollar, *Clypeaster oxybaphon*, which is large and distinctive even in cliff section. The commonest benthic foraminifer is *Lepidocyclina*. Wallace (1969) recognised *Lepidocyclina undosa*, *Heterostegina*, *Miogypsina*, *Dentalina?*, *Bolivina?* and the planktic *Globigerina* from the type area.

Stop 3. Friendship

Drive to Brown's Town and turn left. Travel south on the B3 road. The Friendship pit is about 5 km south of Brown's Town (NGR 115892; Figure 15). This is a worked-out bauxite mine which has largely been infilled. This locality is stratigraphically lower than stop 2, assuming no repetition by faulting.

The limestone is a biosparite, more massively bedded that at stop 2. An excavated bench has yielded over 60 moderately to well-preserved tests of irregular echinoids, particularly a high species of *Clypeaster* sp. nov. very different from *C. oxybaphon*. Other common fossils include colonial scleractinian corals and bivalve molluscs. The mouldic preservation of animals that had an originally aragonitic skeleton is conspicuous. A single shark's tooth has been found at this locality by E.N.D.

Stop 4. Bauxite pipes, Tobolski

Return towards Brown's Town but, on the outskirts, turn southwest towards Tobolski. About halfway between Brown's Town and Tobolski is a large, disused bauxite pit on the left of the road (NGR 109915; Figure 15). Take the track on the left of this pit and park at the limestone quarry. In the adjacent bauxite pit, bauxite infills well-developed solution pipes that are apparent in the face. Note the sharp contact between the bauxite and the Brown's Town Formation. Return to the road and drive towards the next road junction. The exposures on the right just before this point allow close inspection of solution pipes. Vertical tubes of the bivalve *Kuphus* may also be seen in the limestone.

Stop 5. Reclaimed bauxite pit, Tobolski

Continue to drive towards Tobolski, noting the distinctive 'cockpits' developed in the karstified limestone. Shortly before Tobolski (about NGR

Eastern and Central Jamaica

100911; Figure 15) a pit (area 18x10^3 m^2) on the right of the road has been reclaimed by the Kaiser Bauxite Company. This land has been turned over to pasture and conifers. Other reclaimed pits occur in this area, which is still being exploited for bauxite.

ABOVE ROCKS AND BENBOW INLIERS

This excursion examines some of the rocks included within two of the larger Cretaceous inliers that are exposed along the eastern margin of the Clarendon Block (Figure 2). The Above Rocks and Benbow Inliers lie 12 and 32 km north-northwest of Kingston, respectively. These Cretaceous sequences are unconformably overlain by Tertiary limestones of the Yellow Limestone Group and the White Limestone Supergroup.

The oldest rocks of the Above Rocks Inlier belong to the Mount Charles Formation. This unit consists of laminated cherts, mudstones, conglomerates, and andesitic lava flows and volcaniclastics (Reed, 1966). The overlying Border Volcanic Formation outcrops only in the northern half of the inlier and consists of andesite lava flows and volcaniclastics. A hornfels zone marks the contact between the Mount Charles Formation and the intrusive Above Rocks granitoid stock. Although the exact ages of both the Mount Charles and Border Volcanic Formations are still unknown, the granitoid has yielded isotopic dates of 67±5, 64±5 (Chubb & Burke, 1963; see also Casey, 1984) and 60±3.4 Ma (Ahmad *et al.*, 1987b). At least two phases of intrusion are evident, the emplacement of the granitoid stock being followed by the intrusion of numerous silicic dykes (Reed, 1966).

The Benbow Inlier includes some of the oldest dated rocks in Jamaica. The Cretaceous succession in the inlier extends from the Barremian-Aptian Devil's Racecourse Formation, comprising mafic and silicic lava flows and related volcaniclastic rocks, with interbedded, fossiliferous limestone members, through the Aptian-Turonian Rio Nuevo Formation of mudstones and shales, to the volcanic conglomerates of the Tiber Formation (Burke *et al.*, 1968; Robinson *et al.*, 1970; Jiang & Robinson, 1987).

Stop 1. Sue River

Travel north from Kingston on the main A3 road (=Junction Road) (Figure 16). Turn west at Temple Hall and drive through Lawrence Tavern to Glengoffe. Turn right, proceed to Freetown and turn left. Approximately 1 km from Freetown a parochial road on the right leads to the stream.

Eastern and Central Jamaica

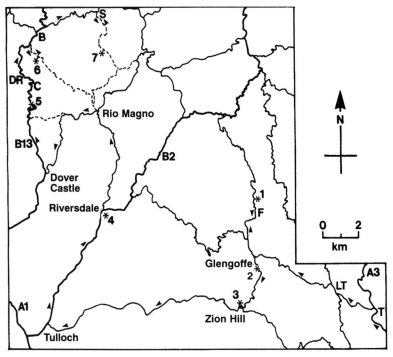

*Figure 16: Outline map of the Above Rocks and Benbow areas.
Key: *=locality (1-7 refer to stops discussed in text); B=Benbow;
C=Copper; DR=Devil's Race Course; F=Freetown; LT=Lawrence
Tavern; S=Springfield; T=Temple Hall. Major roads (A1, A3, B2,
B13) shown as heavy lines, minor roads shown as finer lines, tracks
dashed. Arrows indicate route taken by excursion.*

In the upper reaches of the Sue River, the contact between the granitoid stock and the Mount Charles Formation is exposed (NGR 621678). A pyroxene hornfels zone about 15 m wide passes upstream into a porphyritic fine grained andesite flow (Manning & McCain, 1989). The stock, which is on the downstream side, is strongly altered and mineralised. Fenton (1979) reported the presence of quartz-sericite and quartz-alkali feldspar-biotite alteration assemblages in the granitic rocks. Within these zones of alteration there are disseminated and veined malachite, chalcopyrite and pyrite. Copper concentrations in the soils resting on the stock are as high as 2000 ppm (Fenton, 1979) and conducive for the growth of the fern *Pityrogramma calomelanos*, which is a geobotanical indicator of copper (Burns, 1960).

Eastern and Central Jamaica

Stop 2. Road to Zion Hill bridge

Return to Glengoffe and continue southwards along the road to Zion Hill Bridge. Along this road there are numerous outcrops of silicic dykes intruding the stock (NGR 618650). They are more resistant to weathering than the granites and have been described as granophyres and aplites (Reed, 1966; Manning & McCain, 1989). (A lamprophyre dyke was also identified on the 1993 GA excursion). These authors recognised phenocrysts of plagioclase and alkali feldspars in a crystalline groundmass of alkali feldspar and ferromagnesian minerals. However, a microprobe analysis of one of these dykes, undertaken by T.A.J. and Professor P.W. Scott at the University of Leicester, has shown that the original mineralogy has been lost due to hydrothermal alteration, thus making the original rock difficult to identify.

Stop 3. Zion Hill bridge

Continue south to the Zion Hill bridge, which spans the Rio Pedro (NGR 610626). This is by far the most sampled site of the Above Rocks granitoid, because there are fresh exposures of *in situ* rock in the river bed. At this stop, the stock shows variations in its mineralogy, especially in the proportion of alkali to plagioclase feldspars, so that this outcrop ranges from granite to granodiorite in composition. Other essential minerals are quartz, hornblende and biotite. Accessory minerals include magnetite, apatite, zircon and sphene (Issacs, 1978). Although pyroxene has not been recognised at this stop, it has been detected using microprobe analysis by T.A.J. and Professor P.W. Scott in the granodiorites 2 km northeast of Zion Hill. A similar occurrence of pyroxene was also reported by Isaacs (1978) from the Wind Hill pluton in the Blue Mountain Inlier to the east.

Ellipsoidal-shaped, cognate xenoliths, rich in pyroxene, as well as xenoliths of Mount Charles Formation measuring 1-50 cm in longest dimension, are included within the granitoid. The entire outcrop is penetrated by numerous silicic dykes whose trends are predominantly northwest-southeast and west-northwest-east-southeast (Manning & McCain, 1989).

The Above Rocks stock is calc-alkaline in composition. It is generally more siliceous, with significantly higher K and Rb, than other plutons in Jamaica (Lewis & Gunn, 1972; Jackson *et al.*, 1989). The low initial Sr^{87}/Sr^{86} ratios of 0.703-0.708 for the Above Rocks stock indicates a mantle source, while the strong Light Rare Earth Elements (LREE) normalised pattern for the rock suggests strong fractionation (Jones *et al.*, 1979; Isaacs & Jackson, 1987).

Eastern and Central Jamaica

Stop 4. Riversdale natural bridge

Drive west from Zion Hill to Tulloch. Turn right at Tulloch and proceed north along the B2 to the village of Riversdale (Figure 16).

The abundant outcrops of Tertiary limestones makes Jamaica one of the most appropriate countries in the world for examining the full development of karst topography. At Riversdale one of the few accessible natural bridges on the island is seen (NGR 535677). Large blocks of limestone, that were formerly part of the roof of an underground channel, occur in the river which flows under the bridge.

The limestone that outcrops at this locality is the Upper Eocene Somerset Formation of the White Limestone Supergroup. The rocks display almost horizontal bedding with prominent orthogonal jointing. The Somerset Formation is a generally pink biosparite, rich in foraminifers, benthic molluscs and corals (Wright, 1974), with rare sand dollars.

Stop 5. Berkshire Hall

Drive north to Rio Magno, then turn left (west then southwest) to Dover Castle. At Dover Castle go north to the Devil's Racecourse (=B13 road). At Berkshire Hall (NGR 496731), just south of Devil's Racecourse, there are roadside exposures of the Devil's Racecourse Formation in which there is extensive barytes mineralisation. The vein barytes represents the largest deposit of this mineral in Jamaica. Veins are up to 45 cm wide, trend northwest-southeast, and follow the pattern of faults and tension fractures in the area (Fenton, 1981). The host rocks are hydrothermally-altered, silicic, mainly volcaniclastic rocks containing crystals of feldspar and quartz. These are altered dacites, possibly quartz keratophyres, that show strong island-arc tholeiitic affinities (Jackson, 1987).

Stop 6. Boozy Ridge

Continue north through Copper and Benbow, then take the minor road southeast towards Boozy Ridge. The Benbow Limestone Member of the Devil's Racecourse Formation (Burke *et al.*, 1968; Coates, 1977) is well-exposed along the side of this road (NGR 497757 and adjacent area). Robinson (1994) dated this unit as Lower Aptian, which agrees with the identification from these limestones of the rudist bivalve *Amphitriscoelus waringi* Harris & Hodson by Dr. P.W. Skelton (personal communication). The carbonate horizons within the succession of the Benbow Inlier are typical 'Urgonian' limestones (Ager, 1981), that is, massive, grey limestones with rudist bivalves, associated in this example with nerineacean gastropods and chondrodont bivalves.

Eastern and Central Jamaica

Stop 7. Golden Grove

Return to the B13 road and continue northeast. At Springfield turn south on the minor road. At Golden Grove (NGR 533763) a pillow lava 300 m thick is exposed, which forms the upper part of the Devil's Racecourse Formation (Roobol, 1972). This overlies the 400 m thick Benbow Limestone Member. The rocks display the typical radial jointing and fine grained chilled margins associated with pillow lavas that have cooled rapidly in water. Chlorite and malachite are recognised on the surfaces of several pillows, while calcite and zeolites (natrolite?) infill the joints and vesicles (Burke *et al.*, 1968; Roobol, 1972). Porphyroclasts of albite and sericitized plagioclase feldspar are contained in a fine-grained groundmass of feldspar laths, chlorite, opaque minerals and devitrified glass (Jackson & Smith, 1985).

The altered nature of the pillow lava flow is also evident from its chemistry, which shows high volatile content and the erratic behaviour of Na, K, Rb, Sr and Ba with respect to SiO_2. The fluctuation of these elements, coupled with low grade metamorphic mineralogy and preservation of volcanic texture, suggests that these rocks are basalts that have been hydrothermally altered to spilites. Stable element chemistry confirms that these rocks were originally island arc tholeiitic basalts (Jackson, 1987).

ACKNOWLEDGEMENTS

The authors thank the members of the Geologists' Association who made the 1993 trip to Jamaica, and from which this Guide evolved, for their stimulating company in the field. We also thank David Miller (Department of Geography, UWI), Paul Manning and Steve Kelly (Department of Geology, UWI) for their help in making the original excursion such a success. This guide was significantly improved thanks to the diligence of Trevor Greensmith.

Eastern and Central Jamaica

REFERENCES

(Note: *The Journal of the Geological Society of Jamaica* was formerly called *Geonotes.*)

AGER, D.V. 1981. *The Nature of the Stratigraphical Record.* Macmillan, London.

AHMAD, R., N. LAL & P.K. SHARMA. 1987a. Fission-track age of ignimbrite from Summerfield Formation, Jamaica. *Caribb. J. Sci.*, **23**, 444-448.

------, ------ & ------. 1987b. A fission-track age for the Above Rocks granodiorite, Jamaica. *Caribb. J. Sci.*, **23**, 450-452.

ALLEN, L. & M. NEITA. 1987. Geology of the north Bull Bay sulphate occurrence zone. In (R. Ahmad; ed.) *Proceedings of a Workshop on the Status of Jamaican Geology, Kingston, Jamaica, 14th-16th March, 1984.* Geological Society of Jamaica, Kingston, 282-298.

ANDERSON, R.E. 1971. Notes on the geology of the bauxite deposits of Jamaica. *Spec. Issue J. Geol. Soc. Jamaica, Bauxite/Alumina Symp.*, 9-16.

BANNER, F.T. & W.H. BLOW. 1965. Progress in the foraminiferal biostratigraphy of the Neogene. *Nature*, **208**, 1164-1166.

BATESON, J.H. (compiler). 1974. *Jamaica 1:50,000 Geological Sheet 13 (Provisional) – Mandeville.* Mines and Geology Division, Kingston.

BELL, J. 1986. Caustic waste menaces Jamaica. *New Scient.*, **110**, (1502) 33-37

BLOW, W.H. 1969. Late middle Eocene to Recent planktonic foraminiferal biostratigraphy. In *Proceedings of the First International Conference on Planktonic Microfossils, Geneva, 1967*, 199-422. Brill, Leiden.

BOSS, S.K. & W.D. LIDDELL. 1987. Back-reef and fore-reef analogs in the Pleistocene of north Jamaica: implications for facies recognition and sediment flux in fossil reefs. *Palaios*, **2**, 219-228.

BURKE, K. 1967. The Yallahs Basin: a sedimentary basin southeast of Kingston, Jamaica. *Mar. Geol.*, **5**, 45-60.

------, A.G. COATES & E. ROBINSON. 1968. Geology of the Benbow Inlier and surrounding areas, Jamaica. In (J.B. Saunders; ed.) *Transactions of the Fourth Caribbean Geological Conference, Port of Spain, Trinidad, 28th March-12th April, 1965*, 299-307. Trinidad.

Eastern and Central Jamaica

BURNS, D.J. 1960. Geochemical prospecting for copper in Arthurs Seat, Clarendon, Jamaica. *Geonotes*, **3**, 74-75.

CALDWELL, D.K. 1969. A Miocene needlefish from Bowden, Jamaica. *Q. Jl. Fla. Acad. Sci.*, **28**, 339-344.

CANT, R.V. 1972. Jamaica's Pleistocene reef terraces. *J. Geol. Soc. Jamaica*, **12**, 13-17.

CARBY, B.E. 1985. The geology, mineralogy and geochemistry of the Hope lead-zinc deposit. *Unpubl. Ph.D. thesis, University of the West Indies at Mona.*

CASEY, R. 1964. Granodiorite, Jamaica. In (W.B. Harland, A.G. Smith & B. Wilcox; eds) *The Phanerozoic Time Scale. Supp. Q. Jl. Geol. Soc. Lond.*, **120S**, 284-285.

COATES, A.G. 1977. Jamaican coral-rudist frameworks and their geological setting. In (S.H. Frost, M.P. Weiss & J.B. Saunders; eds) *Reefs and Related Carbonates – Ecology and Sedimentology. Am. Ass. Petrol. Geol., Stud. Geol.*, **4**, 83-91.

CHUBB, L.J. 1960. *The Black Sands of Jamaica*. Unpublished report, Geological Survey Department, Kingston.

------. 1971. Rudists of Jamaica. *Palaeontogr. Am.*, **7**, 157-257.

------ & BURKE, K. 1963. Age of the Jamaican granodiorite. *Geol. Mag.*, **100**, 524-532.

COMER, J.B. 1972. Genesis of Jamaican bauxite. *Unpubl. Ph.D. thesis, University of Texas at Austin.*

DIXON, H.L. & S.K. DONOVAN 1992. The Upper Oligocene Echinoidea of Jamaica. *Geol. Soc. Am. Abstr. w. Progrm.*, **24**, A225-A226.

DONOVAN, S.K. 1993a. Geological excursion guide 9: Jamaica. *Geol. Today*, **9**, 30-34.

------. 1993b. Jamaican Cretaceous Echinoidea. In (R.M. Wright & E. Robinson; eds.) *Biostratigraphy of Jamaica. Geol. Soc. Am. Mem.*, **182**, 93-103.

------ & J.F. BOWEN. 1989. Jamaican Cretaceous Echinoidea. 1. Introduction and reassessment of *?Pygopistes rudistarum* (Hawkins, 1923) n. comb. *Mesozoic Res.*, **2**, 57-65.

Eastern and Central Jamaica

------, H.L. DIXON, R.K. PICKERILL & E.N. DOYLE. 1994a. Pleistocene echinoid (Echinodermata) fauna from southeast Jamaica. *J. Paleont.*, **68**, 351-358.

------ & C.M. GORDON. 1989. Report of a field meeting to selected localities in St Andrew and St Ann, 25 February 1989. *J. Geol. Soc. Jamaica*, **26**, 51-54.

------, ----, C.J. VELTKAMP & A.D. SCOTT. 1993a. Crinoids, asteroids and ophiuroids in the Jamaican fossil record. In (R.M. Wright & E. Robinson; eds) *Biostratigraphy of Jamaica. Geol. Soc. Am. Mem.*, **182**, 842-845.

----, D.A.T. HARPER & E.N. DOYLE. 1993b. A new smooth-shelled *Argyrotheca* Dall (Brachiopoda, Articulata) from the Eocene of Jamaica. *J. Paleont.*, **67**, 1079-1083.

----, T.A. JACKSON & D.T.J. LITTLEWOOD. 1989. Report of a field meeting to the Round Hill region of southern Clarendon, 9 April 1988. *J. Geol. Soc. Jamaica*, **25** (for 1988), 44-47.

----, ---- & R.K. PICKERILL. 1990. Report of a field meeting to selected localities in Portland, 5 May 1990. *J. Geol. Soc. Jamaica*, **27**, 53-57.

---- & D.T.J. LITTLEWOOD. 1993a. The benthic mollusk faunas of two contrasting reef paleosubenvironments: Falmouth Formation (late Pleistocene, last interglacial), Jamaica. *Nautilus*, **107**, 33-42.

---- & ----. 1993b. Late Cenozic reef mollusc faunas of the Caribbean. *Rocks Miner.*, **68**, 226-231.

----, S.A. MILLER, A.P. GRAHAM & H.L. DIXON. 1994b. New fossil crinoids from Jamaica. *J. Paleont.*, **68**, 842-845.

---- & R.K. PICKERILL. 1993. A new species of regular echinoid from the Richmond Formation (Eocene) of Jamaica. *Tertiary Res.*, **14**, 111-115.

---- & C.J. VELTKAMP. 1992. A diadematoid echinoid from the Lower Paleocene of Jamaica. *Caribb. J. Sci.*, **28**, 222-224.

DRAPER, G. 1978. Coaxial pure shear in Jamaican blueschists and deformation associated with subduction. *Nature*, **275**, 735-736.

----. 1987. A revised tectonic model for the evolution of Jamaica. In (R. Ahmad; ed.) *Proceedings of a Workshop on the Status of Jamaican Geology, Kingston, Jamaica, 14th-16th March, 1984*. Geological Society of Jamaica, Kingston, 151-169.

Eastern and Central Jamaica

----, R.R. HARDING, W.T. HORSFIELD, A.W. KEMP & A.E. TRESHAM. 1976. Low-grade metamorphic belt in Jamaica and its tectonic implications. *Geol. Soc. Am. Bull.*, **87**, 1283-1290.

DUNCAN, P.M. & G.P. WALL. 1865. A notice of the geology of Jamaica, especially with reference to the district of Clarendon; with descriptions of the Cretaceous, Eocene and Miocene corals of the island. *Q. Jl. Geol. Soc. Lond.*, **21**, 1-15.

EVA, A & N. McFARLANE. 1985. Tertiary to early Quaternary carbonate facies relationships in Jamaica. In (editor unknown) *Transactions of the Fourth Latin American Geological Congress, Port-of-Spain, Trinidad, 7th-15th July, 1979*, **1**, 210-219. Trinidad.

FENTON, A. 1979. Copper prospects of Jamaica; a geological review. *Bull. Geol. Surv. Div., Kingston*, **9**, 225 pp.

----. (ed.). 1981. The mineral resources of Jamaica. *Bull. Geol. Surv. Div., Kingston*, **8**, 104 pp.

GILI, E., J.P. MASSE & P.W. SKELTON. 1990. Did rudists build reefs? *Abstracts, Second International Conference on Rudists, October, Università "La Sapienza", Rome, and Università degli Studi, Bari*, p. 9

GEOLOGICAL SURVEY OF JAMAICA. 1958. *Jamaica, Geology. Scale 1:250,000.* Directorate of Overseas Surveys, London.

GOODFRIEND, G.A. & R.M. MITTERER. 1988. Late Quaternary land snails from the north coast of Jamaica: local extinction and climatic change. *Palaeogeogr., Palaeoclimat., Palaeoecol.*, **63**, 293-311.

GORDON, C.M. & S.K. DONOVAN. 1992. Disarticulated echinoid ossicles in paleoecology and taphonomy: the last interglacial Falmouth Formation of Jamaica. *Palaios*, **7**, 157-166.

GREEN, G.W. (compiler). 1974. *Kingston. Geological Sheet 25. 1:50,000 Jamaica.* Ministry of Mining and Natural Resources, Kingston.

----. 1977. Structure and stratigraphy of the Wagwater Belt, Kingston, Jamaica. *Overseas Geol. Miner. Resour.*, **48**, 1-21.

GRIPPI, J. 1980. Geology of the Lucea Inlier, western Jamaica. *J. Geol. Soc. Jamaica*, **19**, 1-24.

HARLAND, W.B., R.L. ARMSTRONG, A.V. COX, L.E. CRAIG, A.G. SMITH & D.G. SMITH. 1990. *A Geologic Time Scale 1989.* Cambridge University Press, Cambridge.

Eastern and Central Jamaica

HARPER, D.A.T. & S.K. DONOVAN. 1990. Fossil brachiopods of Jamaica. *J. Geol. Soc. Jamaica*, **27**, 27-32.

HAWKINS, H.L. 1927. Descriptions of new species of Cainozoic Echinoidea from Jamaica. *Mem. Mus. Comp. Zool. Harv.*, **50**, 76-84.

HILL, V.G. 1955. The mineralogy and genesis of bauxite deposits of Jamaica, B.W.I. *Am. Miner.*, **40**, 676-688.

HOLLIDAY, D.W. 1971. Origin of Lower Eocene gypsum-anhydrite rocks, southeast St. Andrew, Jamaica. *Trans. Inst. Min. Metall.*, **B80**, 305-315.

HORSFIELD, W.T. 1974. Major faults in Jamaica. *J. Geol. Soc. Jamaica*, **14**, 1-14.

HOSE, H.R. & H.R. VERSEY. 1957. Palaeontological and lithological divisions of the lower Tertiary limestones of Jamaica. *Colon. Geol. Miner. Resour.*, **6** (for 1956), 19-39.

ISAACS, M.C. 1978. A petrological study of plutonic rocks from central and eastern Jamaica. *Unpubl. M.Sc. thesis, University of Oxford*.

---- & T.A. JACKSON. 1987. The mineralogy and geochemistry of plutonic rocks from Jamaica. In (R. Ahmad; ed.) *Proceedings of a Workshop on the Status of Jamaican Geology, Kingston, Jamaica, 14th-16th March, 1984*. Geological Society of Jamaica, Kingston, 95-106.

JACKSON, T.A. 1977. The petrochemistry and origin of the Tertiary volcanic rocks, Wagwater Belt, Jamaica, W.I. *Unpubl. Ph.D. thesis, University of the West Indies, Mona*.

----. 1986. St. Peter's Inlier – fact or fiction. *J. Geol. Soc. Jamaica*, **23** (for 1985), 44-49.

----. 1987. The petrology of Jamaican Cretaceous and Tertiary volcanic rocks and their tectonic significance. In (R. Ahmad; ed.) *Proceedings of a Workshop on the Status of Jamaican Geology, Kingston, Jamaica, 14th-16th March, 1984*. Geological Society of Jamaica, Kingston, 107-119.

---- & T.E. SMITH. 1978. Metasomatism in the Tertiary volcanics of the Wagwater Belt, Jamaica. *Geologie Mijnb.*, **57**, 212-220.

---- & ----. 1979. The tectonic significance of basalts and dacites in the Wagwater Belt, Jamaica. *Geol. Mag.*, **116**, 365-374.

Eastern and Central Jamaica

---- & ----. 1982. Mesozoic and Cenozoic mafic magma types of Jamaica and their tectonic setting. In (W. Snow, N. Gil, R. Llinas, R. Rodrigues-Torres, M. Seaward & I Tavares; eds) *Transactions of the Ninth Caribbean Geological Conference, Santo Domingo, Dominican Republic, 16th-20th August, 1980*, 2, 435-440.

---- & ----. 1985. The petrochemistry of some Cretaceous mafic volcanics, Jamaica, West Indies. In *Transactions of the Fourth Latin American Geological Congress, Port-of-Spain, Trinidad, 7th-15th July, 1979*, 1, 387-396.

----, ---- & M.C. ISAACS. 1989. The significance of geochemical variations in Cretaceous volcanic and plutonic rocks of intermediate and felsic composition from Jamaica. *J. Geol. Soc. Jamaica*, 26, 33-42.

JIANG, M.-J & E. ROBINSON. 1987. Calcareous nannofossils and larger Foraminifera in Jamaican rocks of Cretaceous to early Eocene age. In (R. Ahmad; ed.) *Proceedings of a Workshop on the Status of Jamaican Geology. Spec. Pub. Geol. Soc. Jamaica*, 24-51.

JONES, L.M., R.L. WALKER, S.E. KESLER & J.F. LEWIS. 1979. Strontium isotope geochemistry of late Cretaceous granodiorites, Jamaica and Haiti, Greater Antilles. *Earth Planet. Sci. Lett.*, 43, 112-116.

KAUFFMAN, E.G. & C.C. JOHNSON. 1988. The morphological and ecological evolution of Middle and Upper Cretaceous reef-building rudistids. *Palaios*, 3, 194-216.

---- & N.F. SOHL. 1974. Structure and evolution of Antillean Cretaceous rudist frameworks. *Verh. Naturf. Ges. Basel*, 84, 399-467.

KESLER, S.E., E. LEVY & C. MARTIN F. 1990. Metallogenic evolution of the Caribbean region. In (G. Dengo & J.E. Case; eds) *The Geology of North America. Volume H. The Caribbean Region*, 459-482. Geological Society of America, Boulder.

LAGAAIJ, R. 1959. Some species of Bryozoa new to the Bowden Beds, Jamaica, B.W.I. *Micropaleontology*, 5, 482-486.

LAND, L.S. 1973. Holocene meteoric dolomitization of Pleistocene limestones, north Jamaica. *Sedimentology*, 20, 411-424.

----. 1991. Some aspects of the late Cenozoic evolution of north Jamaica as revealed by strontium isotope stratigraphy. *J. Geol. Soc. Jamaica*, 28, 45-48.

---- & S. EPSTEIN. 1970. Late Pleistocene diagenesis and dolomitization, north Jamaica. *Sedimentology*, **14**, 187-200.

LARSON, D.C. 1983. Depositional facies and diagenetic fabrics in the late Pleistocene Falmouth Formation of Jamaica. *Unpubl. M.S. thesis, University of Oklahoma, Norman.*

LEWIS, J.F. & B.M. GUNN. 1972. Aspects of island arc evolution and magmatism in the Caribbean: geochemistry of some West Indian plutonic and volcanic rocks. In (C. Petzall; ed.) *Transactions of the Sixth Caribbean Geological Conference, Margarita, Venezuela, 6th-14th July, 1971*, 171-177.

LIDDELL, W.D., S.L. OHLHORST & A.G. COATES. 1984. *Modern and Ancient Carbonate Environments of Jamaica. Sedimenta*, **10**, 98 pp. Rosenstiel School of Marine and Atmospheric Science, University of Miami, Miami.

LITTLEWOOD, D.T.J. & S.K. DONOVAN. 1988. Variation of Recent and fossil *Crassostrea* in Jamaica. *Palaeontology*, **31**, 1013-1028.

MACPHEE, R.D.E., D.C. FORD & D.A. McFARLANE. 1989. Pre-Wisconsinan mammals from Jamaica and models of late Quaternary extinction in the Greater Antilles. *Quaternary Res.*, **31**, 94-106.

MAHARAJ, R.J. 1990. 'Landslides' in the parish of St. Andrew, Jamaica: report of a field meeting on the Irish Town Road, Junction Road and St. Joseph Road, Kintyre, 13 May 1989. *J. Geol. Soc. Jamaica*, **27**, 45-51.

MANN, P. & K. BURKE. 1990. Transverse intra-arc rifting: Palaeogene Wagwater Belt, Jamaica. *Mar. Petrol. Geol.*, **7**, 410-427.

----, G. DRAPER & K. BURKE. 1985. Neotectonics and sedimentation at a strike-slip restraining bend, Jamaica. In (K.T. Biddle & N. Christie-Blick; eds.) *Strike-Slip Deformation, Basin Formation, and Sedimentation. Spec. Publ. SEPM*, No. **37**, 211-226.

MANNING, P.A.S. & T. McCAIN. 1989. Report of a field meeting to the Above Rocks Inlier, north St. Catherine and south St. Mary, 3 December 1988. *J. Geol. Soc. Jamaica*, **26**, 43-50.

----, ---- & R. AHMAD. 1992. Landslides triggered by 1988 Hurricane Gilbert along roads in the Above Rocks area, Jamaica. In (R. Ahmad; ed.) *Natural hazards in the Caribbean. Spec. Issue J. Geol. Soc. Jamaica*, **12**, 34-53.

Eastern and Central Jamaica

MATLEY, C.A. 1951. *Geology and Physiography of the Kingston District, Jamaica.* Crown Agents, London.

McFARLANE, N. (compiler). 1977a. *Jamaica – Geology 1:250,000 sheet.* Mines and Geology Division, Kingston.

----. 1977b. The non-carbonate Pleistocene sand deposits of the south central coast of Jamaica. *Abstract, Tenth INQUA Conference, University of Birmingham, August 1977.*

MORRIS, S.F. 1993. The fossil arthropods of Jamaica. In (R.M. Wright & E. Robinson; eds.) *Biostratigraphy of Jamaica. Geol. Soc. Am. Mem.*, **182,** 115-124.

MULLINS, H.T., A.C. NEUMANN, R.J. WILBER & M.R. BOARDMAN. 1980. Nodular carbonate sediment on Bahamian slopes: possible precursors to nodular limestones. *J. Sedim. Petrol.*, **50,** 117-131.

PALMER, D.K. 1945. Notes on the Foraminifera from Bowden, Jamaica. *Bull. Am. Paleont.*, **29,** 78 pp.

PICKERILL, R.K. & S.K. DONOVAN. 1991. Observations on the ichnology of the Richmond Formation of eastern Jamaica. *J. Geol. Soc. Jamaica*, **28,** 19-35.

----, ---- & H.L. DIXON. 1992. The Richmond Formation of eastern Jamaica revisited – further ichnological observations. *Caribb. J. Sci.*, **28,** 89-98.

----, ----, & ----. 1993a. The trace fossil *Dactyloidites ottoi* (Geinitz, 1849) from the Neogene August Town Formation of south central Jamaica. *J. Paleont.*, **67,** 1070-1074.

----, ----, ---- & E.N. DOYLE. 1993b. Ichnology of the Paleogene Richmond Formation of eastern Jamaica – the final chapter? *Atlant. Geol.*, **29,** 61-67.

----, ----, ---- & ----. 1993c. *Bichordites monastiriensis* from the Pleistocene of southeast Jamaica. *Ichnos*, **2,** 225-230.

PORTER, A.R.D., T.A. JACKSON & E. ROBINSON. 1982. *Minerals and Rocks of Jamaica.* Jamaica Publishing House, Kingston.

PRESCOTT, G.C. & H.R. VERSEY. 1958. Field meeting at Hayes Common and Round Hill, Jamaica. *Proc. Geol. Ass.*, **69,** 38-39.

Eastern and Central Jamaica

RÁCZ, L. 1971. Two new Pliocene species of *Neomeris* (calcareous algae) from the Bowden Beds, Jamaica. *Palaeontology*, **14**, 623-628.

REED, A.J. 1966. Geology of the Bog Walk Quadrangle, Jamaica. *Bull. Geol. Surv. Dept., Kingston*, **6**, 54 pp.

ROBINSON, E. 1958a. The younger rocks of St. James and Trelawny. *Geonotes*, **1**, 15-17.

----. 1958b. The Buff Bay Beds and Low Layton volcanics. *Geonotes*, **1**, 66-71.

----. 1959. Field meeting at Buff Bay and Low Layton, Jamaica. *Proc. Geol. Ass.*, **70**, 271-272.

----. 1965. Tertiary rocks of the Yallahs area, Jamaica. *J. Geol. Soc. Jamaica*, **7**, 18-27.

----. 1967a. Biostratigraphic position of late Cainozoic rocks in Jamaica. *J. Geol. Soc. Jamaica*, **9**, 32-41.

----. 1967b. Submarine slides in White Limestone Group, Jamaica. *Am. Ass. Petrol. Geol. Bull.*, **51**, 569-578.

----. 1968. The geology of Round Hill, Clarendon. *J. Geol. Soc. Jamaica*, **9** (for 1967), 46-47.

----. 1969. Geological field guide to Neogene sections in Jamaica West Indies. *J. Geol. Soc. Jamaica*, **10**, 1-24.

----. 1971. Observations on the geology of Jamaican bauxite. *Spec. Issue J. Geol. Soc. Jamaica, Bauxite/Alumina Symp.*, 3-9.

----. 1988. Late Cretaceous and early Tertiary sedimentary rocks of the Central Inlier, Jamaica. *J. Geol. Soc. Jamaica*, **24** (for 1987), 49-67.

----. 1994. Jamaica. In (S.K. Donovan & T.A. Jackson; eds.) *Caribbean Geology: An Introduction*. University of the West Indies Publishers' Association, Kingston, 111-127.

---- & M.M.-J. JIANG. 1990. Paleogene calcareous nannofossils from western Portland, and the ages and significance of the Richmond and Mooretown Formations of Jamaica. *J. Geol. Soc. Jamaica*, **27**, 17-25.

----, J.F. LEWIS & R. CANT. 1970. Field guide to aspects of the geology of Jamaica. In (T.W. Donnelly; ed.) *International Field Institute Guidebook to the Caribbean Island Arc System*. American Geological Institute, Washington, D.C., 48 pp.

Eastern and Central Jamaica

ROOBOL, M.J. 1972. Volcanic geology of Jamaica. In (C. Petzall; ed.) *Transactions of the Sixth Caribbean Geological Conference, Margarita Island, Venezuela, 6th-14th July, 1971*, 100-107. Escuela de Geologia y Minas, Caracas.

----. 1976. Post-eruptive mechanical sorting of pyroclastic material – an example from Jamaica. *Geol. Mag.*, **113**, 429-440.

---- & W.T. HORSFIELD. 1976. Sea floor lava outcrop in the Jamaica passage. *J. Geol. Soc. Jamaica*, **15**, 7-10.

ROYALL, M.T. & J. BANHAM. 1981. A review of the thermal and mineral springs of Jamaica. In (A. Lyew-Ayee; ed.) *Proceedings of an Industrial Minerals Symposium, Kingston, 21-25 September, 1981. Special Issue of the Journal of the Geological Society of Jamaica*, 83-93.

SAVAGE, R.J.G. 1990. Preliminary report on the vertebrate fauna of the Red Hills Road cave, St. Andrew, Jamaica. *J. Geol. Soc. Jamaica*, **27**, 33-35.

SAWKINS, J.G. 1869. Report on the geology of Jamaica. *Mem. Geol. Surv.*, London, 339 pp.

SCHMIDT, W. 1988. Stratigraphy and depositional environment of the Lucea Inlier, western Jamaica. *J. Geol. Soc. Jamaica*, **24** (for 1987), 15-35.

SCOFFIN, T.P. 1987. *An Introduction to Carbonate Sediments and Rocks*. Blackie, Glasgow, 274 pp.

SCOTT, P.W. & Y.D. DRAKAPOULOS. 1989. An occurrence of barytes in the Wagwater Group near Irish Town, St. Andrew. *J. Geol. Soc. Jamaica*, **26**, 19-21.

SINCLAIR, I.G. 1966. Estimation of the thickness required to produce the Jamaican bauxites by the residual process. *J. Geol. Soc. Jamaica*, **8**, 24-31.

SKELTON, P.W. 1985. Preadaptation and evolutionary innovation in rudist bivalves. In (J.C.W. Cope & P.W. Skelton; eds.) *Evolutionary Case Histories from the Fossil Record. Spec. Pap. Palaeont.*, **33**, 159-173.

----. 1991. Morphogenetic versus environmental cues for adaptive radiations. In (N. Schmidt-Kittler & K. Vogel; eds.) *Constructional Morphology and Evolution*. Springer-Verlag, Berlin, 375-388.

----, S.K. DONOVAN & H.L. DIXON. 1992. Palaeoecology of the giant Antillean rudist bivalve *Titanosarcolites giganteus* (Whitfield). *Paleont. Newsl.*, **16**, 20.

Eastern and Central Jamaica

SOHL, N.F. 1992. Upper Cretaceous gastropods (Fissurellidae, Haliotidae, Scissurellidae) from Puerto Rico and Jamaica. *J. Paleont.*, **66**, 414-434.

STERN, R.J., P.N. LIN, J.D. MORRIS, M.C. JACKSON, P. FRYER, S.H. BLOOMER & E. ITO. 1990. Enriched back-arc basin basalts from the northern Mariana Trough: implications for the magmatic evolution of back-arc basins. *Earth Planet. Sci. Lett.*, **100**, 210-225.

TRECHMANN, C.T. 1924. The Carbonaceous Shale or Richmond Formation of Jamaica. *Geol. Mag.*, **61**, 2-19.

----. 1930. The Manchioneal Beds of Jamaica. *Geol. Mag.*, **67**, 199-218.

VERSEY, H.R. 1959. Recent work on the Milk River mineral spring. *Geonotes*, **2**, 123-128.

VINCENZ, S.A. 1959. Some observations of gamma radiation emitted by a mineral spring in Jamaica. *Geophys. Prospect.*, **7**, 433-434.

WADGE, G. 1982. A Miocene submarine volcano at Low Layton, Jamaica. *Geol. Mag.*, **119**, 193-199.

----, T.A. JACKSON, M.C. ISAACS & T.E. SMITH. 1982. The ophiolitic Bath-Dunrobin Formation, Jamaica: significance for Cretaceous plate margin evolution in the north-western Caribbean. *J. Geol. Soc. Lond.*, **139**, 321-333.

WALLACE, R.J. 1969. The paleoecology of the Browns Town and Montpelier Limestones (Oligocene-Miocene) of Jamaica. *Unpubl. M.S. thesis, Northern Illinois University, Dekalb.*

WESCOTT, W.A. & F.G. ETHRIDGE. 1980. Fan-delta sedimentology and tectonic setting – Yallahs fan delta, southeast Jamaica. *Am. Ass. Petrol. Geol. Bull.*, **64**, 374-399.

---- & ----. 1983. Eocene fan delta-submarine fan deposition in the Wagwater Trough, east-central Jamaica. *Sedimentology*, **30**, 235-247.

WOOD, P.A. 1976. Beaches of accretion and progradation in Jamaica. *J. Geol. Soc. Jamaica*, **15**, 24-31.

WOODLEY, J.D. & E. ROBINSON. 1977. *Field Guidebook to the Modern and Ancient Reefs of Jamaica. Third International Coral Reef Symposium.* Rosenstiel School of Marine and Atmospheric Science, University of Miami, Miami.

Eastern and Central Jamaica

WOODRING, W.P. 1925. Miocene mollusks from Bowden, Jamaica. *Publs Carnegie Instn.*, **366**, 222 pp.

----. 1928 Miocene mollusks from Bowden, Jamaica, pt. II, gastropods and discussion of results. *Publs Carnegie Instn.*, **385**, 564 pp.

WRIGHT, R.M. (ed.). 1974. Field guide to selected Jamaican geological localities. *Spec. Publ. Mines Geol. Div., Kingston*, **1**, 57 pp.

ZANS, V.A. 1958. Recent views on the origin of bauxite. *Geonotes*, **1**, 123-132.

----. 1959. Judgement Cliff landslide in the Yallahs Valley. *Geonotes*, **2**, 43-48.

----. 1960. 'God's Well' and its origin. *Geonotes*, **3**, 98-105.

Eastern and Central Jamaica

GLOSSARY

This glossary explains some of the more specialised terms referred to in the text.

AMPHIBOLITE. An amphibole-rich metamorphic rock.

APHANITIC. A fine grained (<1 mm) volcanic rock.

APHYRIC. Refers to igneous rocks that are non-porphyritic, that is, that lack phenocrysts.

BIOMICRITE. A biomicritic limestone is composed of fossil remains in a matrix of mud-sized carbonate grains.

BLUESCHISTS. Schistose metamorphic rocks containing blue amphiboles, such as glaucophane.

CALICHE. A crust of calcite, produced by precipitation during evaporation of water in areas of low rainfall.

CATACLASIS. Deformation of a rock produced by faulting, involving the fracture and rotation of mineral grains or grain aggregates without chemical reconstitution.

COGNATE XENOLITHS. Inclusions within an igneous rock to which the rock is genetically related.

CUPROPHILIC. Literally 'copper loving', as in some plants that prefer to grow on soils rich in copper.

DACITE. A quartz-bearing volcanic rock composed predominantly of plagioclase feldspar, hornblende and biotite.

DRIPSTONE. Calcitic deposits produced by precipitation from dripping water, rich in dissolved calcium carbonate; usually formed in caves within limestones.

EPITHERMAL MINERALISATION. Hydrothermal mineralization that takes place within the 50°-200°C temperature range

EUSTASY. Global changes in sea-level.

FAN DELTA. An alluvial fan with an associated, more distal delta, found in regions where highland areas are close to the coast.

Eastern and Central Jamaica

FLYSCH. Interbedded finer (mudrocks, siltstones) and coarser (generally sandstones) grained, clastic sedimentary rocks, deposited under deeper water, turbiditic conditions and usually associated with rapid uplift of the sediment source area.

GRANODIORITE. A leucocratic, igneous, plutonic rock containing plagioclase feldspar, potassium feldspar and quartz, with biotite, hornblende and occasionally pyroxene as mafic components.

GREENSCHIST. A schistose metamorphic rock containing green minerals, such as chlorite, epidote and actinolite.

HYALOCLASTITES. Glassy volcaniclastic rocks produced by reaction between magma and sea water (or ice).

ICHNOCOENOSIS. An association of trace fossils preserved in a specific sedimentary environment and produced by a particular community of benthic organisms.

ICHNOLOGY. The study of trace fossils.

ICHNOTAXA. A named trace fossil. Although trace fossils are sedimentary structures (albeit biologically generated), they are given Latinised binomens as if they were truly organisms.

KARST. Topography associated with limestone (or other soluble rock), generally in regions of at least moderate rainfall, and produced by solution which occurs mainly in the subsurface. Sink holes are one type of feature.

KERATOPHYRES. Alkali-rich igneous rocks displaying a volcanic texture, and containing albite, chlorite and epidote.

LEUCOCRATIC. Light-coloured igneous rocks containing less than 30% dark minerals.

MEGACRYSTS. Very large crystals contained in the finer-grained groundmass of an igneous or metamorphic rock.

MONOMICT CONGLOMERATE. A conglomerate in which the clasts are composed of one rock type only.

POLYMICT CONGLOMERATE. A conglomerate in which the clasts are composed of more than one rock type.

PORPHYRY. An igneous rock that contains phenocrysts in a finer-grained groundmass.

PORPHYROCLASTS. Original crystals within a fine-grained, altered groundmass.

SILICICLASTICS. Sediments formed predominantly of detrital grains of quartz, feldspar and mica.

SPILITES. An altered basaltic rock displaying a volcanic texture, and containing albite, chlorite, epidote and/or prehnite and actinolite.

TEPEE STRUCTURES. Desiccation structures produced in carbonate environments and having a triangular (tepee-like) cross-section.

TERRACE. A flat-lying area of land in a region of stepped topography, with a steep descending slope on its more seawards margin and a steep ascending slope on its more landwards margin (these steep slopes may be cliffs or ancient cliff-lines).

TOMBOLO. An elongate spit of unconsolidated sediment that connects an island to the adjacent coast, often produced by longshore drift.

TRACE FOSSILS. Sedimentary structures produced by the actions of organisms.

TRANSPRESSION. Compression associated with strike-slip faulting during deformation.

TRANSTENSION. Extension associated with strike-slip faulting during deformation.

UNROOFING. Surface exposure of a pluton following intrusion, cooling and erosion of the overlying country rock.

WACKESTONE. A mud-supported limestone in which carbonate clasts form over 10% of the rock.